OXIDES AND OXIDE FILMS

Volume 5

Edited by

Ashok K. Vijh

Hydro-Quebec Institute of Research
Varennes, Quebec, Canada

MARCEL DEKKER, INC. New York and Basel

Library of Congress Cataloging in Publication Data (Revised)

Diggle, John W
Oxides and oxide films.

(The Anodic behavior of metals and semiconductors series)
Vol. : edited by J. W. Diggle and A. K. Vijh.
Vol. edited by A.K. Vijh.
Includes bibliographies.
1. Metallic oxides. 2. Metallic films. 3. Metals--Anodic oxidation. II. Vijh, Ashok K. II. Title.
III. Series.
QD561.D54 542'.721'2 72-83120
ISBN *0-8247-6580-X*

MARCEL DEKKER, INC.
270 Madison Avenue, New York, New York 10016

Current printing (last digit):
10 9 8 7 6 5 4 3 2 1

PRINTED IN THE UNITED STATES OF AMERICA

ABOUT THE SERIES

The aim of this series of books as originally conceived was to treat all aspects of the anodic behavior of materials. The subjects related to this general theme are anodic oxide films, bulk oxide materials, corrosion reaction layers, anodic dissolution of metals, anodic oxidation of organics, and the anodic gas evolution reactions. Since the inception of this series, several volumes (e.g., on organic electrochemistry and corrosion science) have appeared containing chapters on most of the foregoing topics, with the conspicuous exception of oxides and oxide films. However, in the last decade or so, an increasingly large amount of research activity and technological interest has been evident in the areas of oxidation of metals, oxide films, and oxides; these subjects are treated as a unified theme only in this series. In order to avoid duplication in other areas and to provide a prominent forum for authoritative chapters to the ever-expanding community of research investigators working on oxides and oxide films, it appears appropriate now to concentrate this series on matters related only to oxides. Oxides and oxide films cover a very wide spectrum of interdisciplinary research effort having important bearings on modern technology. The modified scope of the series would thus include topics such as anodic, thermal, and plasma oxide growth on metals, alloys, and semiconductors; surface specificity and reactivity of oxides and oxide films; electrocatalysis and catalysis on oxides; oxide layers in corrosion; anodic finishing of metals, i.e., anodizing; oxide ceramics; solid oxide electrolytes; nonstoichiometry, defect structure, and diffusion in oxides; mechanisms of electronic and ionic conduction; dielectric properties; rectifying characteristics; electrical breakdown; electroluminescence; mechanical properties; tribological (e.g., hardness, friction, and wear) characteristics; magnetic properties; optical behavior; oxide glasses; colloidal and interfacial characteristics of oxide suspensions in aqueous and nonaqueous media; electrical barriers and epitaxial problems at metal-oxide interfaces; chemical and electrochemical reactions of oxides such as their dissolution and hydration; new techniques for the measurement of properties of oxides and oxide films; and, finally, oxides in industrial products and processes such as batteries, capacitors, MOS and related technologies, dielectric and insulating materials, thin film devices, nuclear technology, energy conversion and storage system, etc.

The above list of topics is not intended to be exclusive but only indicates some typical subjects of relevance to the series. All important aspects of the oxidation of metals, alloys, and semiconductors as well as the salient properties of oxides, oxide films, and oxidized surfaces would fall within the scope of this series.

This series of volumes should thus be of interest to scientists, engineers, and technologists whose main work touches one of the following disciplines: materials science; applied physics; electrical engineering; electrochemistry; solid state chemistry; corrosion science and engineering; heterogeneous catalysis; surface science and technology; and the various fields of industrial endeavor in which oxides and oxide films play a significant role.

The Editor encourages the readers to present their opinions and comments on the modified definition of the scope of this series, as well as any suggestions regarding the contents, orientation, and the possible contributors to the future volumes. This invitation is extended in the belief that comments and feedback only from the book reviewers and the personal acquaintances of the Editor are never quite sufficient for creating volumes that completely meet the real needs of a wide variety of readers.

PREFACE

This volume consists of two chapters, each of which deals with the formation and properties of anodic oxides on a series of metals. These two chapters are complementary in nature in that one describes the formation of anodic oxides in liquid electrolytes whereas the other examines the cases in which the liquid electrolyte has been replaced by some form of oxygen discharge in the gas phase.

The first chapter is a review of the recent studies on the anodic oxide growth on noble metals and their alloys. This work constitutes a major theme in modern electrochemical surface science, owing both to its intrinsic fundamental importance and to its possible applications in devices in which good electrocatalytic surfaces are essential.

Chapter 2 describes the techniques for producing gaseous electrolytes and the formation of anodic oxides carried out in them. The properties of the oxides thus formed and their growth mechanisms are also discussed. The oxides formed by gas discharge anodization can find diverse applications as, e.g., in capacitors, Josephson junctions, Xerography and planar transistors.

The Editor is indebted to the late John W. Diggle who was involved in the initial planning of this book but whose untimely death deprived us of his participation as the co-editor of this volume.

Grateful acknowledgement is made to Dr. L. Boulet, F.R.S.C., the founding Director of the Hydro-Quebec Institute of Research (IREQ) for providing atmosphere, facilities and inspiration conducive to the pursuit of scientific work. Dr. G. G. Cloutier, F.R.S.C., Assistant Director of IREQ is thanked for his active interest and personal encouragement in the Editor's work. I am also indebted to Dr. R. Bartnikas, Fellow IEEE, Scientific Director, Materials Science Department, IREQ, for his enlightened appreciation of my scientific and technical endeavors.

The Editor's greatest and very personal gratitude is due to his wife Danielle and his son Aldous. Their comforting presence and willingness to sacrifice many hours that in fact belonged to them were indispensable to the execution of the present project.

Ashok K. Vijh

CONTRIBUTORS TO VOLUME 5

GUY BELANGER, Hydro-Quebec Institute of Research, Varennes, Quebec, Canada

JOHN F. O'HANLON, IBM Corporation, Thomas J. Watson Research Center, Yorktown Heights, New York

ASHOK K. VIJH, Hydro-Quebec Institute of Research, Varennes, Quebec, Canada

CONTENTS OF VOLUME 5

CONTENTS OF VOLUME 5

CONTENTS OF OTHER VOLUMES

OXIDES AND OXIDE FILMS

Volume 5

Chapter 1

ANODIC OXIDES ON NOBLE METALS

Guy Bélanger and Ashok K. Vijh

Hydro-Quebec Institute of Research
Varennes, Quebec, Canada

I. INTRODUCTION AND SCOPE OF THE REVIEW

The object of this chapter is to review the work on the anodic oxide formation on noble metals. The metals to be covered are platinum, palladium, rhodium, iridium, osmium, ruthenium, gold, and their alloys. For the purposes of the review, silver is not considered as a noble metal mainly because on anodization it forms a large variety of oxides and other corrosion products and, in this sense, its anodic behavior approaches that of other nonnoble metals such as nickel, copper, cobalt, and lead.

Among the noble metals, platinum is the most thoroughly examined as regards anodic oxide formation and reduction, undoubtedly owing to the unique position of this metal as an excellent electrocatalyst in electrode reactions. Several previous reviews [1,2] on the anodic oxide formation on platinum and other noble metals exist in the literature and are still of considerable value. This chapter aims to cover more recent literature, although salient points brough out by representative older publications [1-12] are also outlined. That a new up-to-date review on the subject is indeed timely is suggested not only by an enormous recent activity [13-35] in the field but also by the fact that several important nonelectrochemical techniques, such as ellipsometry, X-ray photoelectron spectroscopy (low energy electron diffraction, electron spectroscopy for chemical analysis), Auger spectroscopy, and specular reflection spectroscopy, have been widely applied to the problem of anodic oxidation of noble metals only since the late 1960s.

II. ANODIC OXIDATION OF PLATINUM

A. Introduction and Summary of the Work up to 1966

Of all the noble metals, platinum has been most widely studied as regards its anodic oxidation behavior. The practical importance of platinum as an electrocatalyst and its several other attractive properties, such as its corrosion resistance, make it an obvious choice for extensive investigation. The fact that reproducible platinum surfaces can be obtained by appropriate treatments, and their inert behavior in acidic and alkaline media, imparts to platinum a unique role in electrocatalysis, electroanalysis, and other related areas of research.

In the following sections, attention will be paid to the anodic oxide formation and reduction on platinum and the related mechanistic aspects. The oxygen reduction reaction, the oxygen evolution reaction, and the open-circuit behavior of platinum in oxygen-saturated solutions will not be treated as such but may be touched upon only insofar as they bear upon the problem of the oxidation of platinum; these topics were reviewed by Hoare [1] in 1967. Gilman [2] has reviewed the literature published prior to 1966. In that review, the topics treated were the charge balance between the anodic oxide film formation and its reduction, the stoichiometry of this oxide, the kinetics of anodic oxide growth, and some other related subjects.

The determination of the oxygen-platinum ratio (i.e., the stoichiometry of the oxide) is usually carried out electrochemically by measuring the charge needed to form the oxide, Q_O, and then comparing this charge with the one associated with the hydrogen atom adsorption on the metal, ${}^{S}Q_H$, assuming that each metal site adsorbs one hydrogen atom. The ratio of Q_O and ${}^{S}Q_H$ has been reported to be 2 [2,3] corresponding to the oxide PtO. The same result was also obtained by Will and Knorr [4]. This represents the oxygen film formation prior to the oxygen evolution as measured in the potentiodynamic profiles. The deviations from $Q_O / {}^{S}Q_H = 2$ were attributed to the presence of traces of organic impurities in solution which would tend to increase the Q_O value [2].

The growth of the oxide on platinum can be followed as a function of potential by integration of the current-time curve at a given formation potential [2]. No limiting coverage was found even for potentials up to 2 volts (V) [2]; this behavior was also observed by Becker and Breiter [5], Laitinen and Enke [6], and Visscher and Devanathan [7]. Gilman [2], on the other hand, concluded that the coverage does not exceed a monolayer of oxygen, i.e., the ratio $Q_O / {}^SQ_H = 2$ at about 1.4 V.

The discrepancies found between the anodic oxide charge formation and the cathodic charge associated with the oxide reduction have also been discussed [2]. Several workers found an equality between Q_O (anodic) and Q_O (cathodic) [3,4,8] for potentials not exceeding 1.6 V. The ratios of Q_O approximately equal to 2 were reported by Dietz and Göhr [9] and Vetter and Berndt [10] on the basis of galvanostatic charging curves. Gilman [2] attributed these discrepancies to impurity effects since the time involved in galvanostatic charging curves at low current density is rather long and consequently favors the complication due to impurities.

The capacity-potential relationships were also reviewed by Gilman [3]; a maximum in capacity is observed at approximately 0.9 V and this maximum was interpreted as an indication of the reversible reaction

$$Pt + \rightleftarrows PtOH + H^+ + e \tag{1}$$

As the anodic potential increases further, the capacity of the smooth platinum electrode decreases indicating the irreversibility of the oxide formation process.

The nature of the anodic film on platinum has been a subject of much controversy: is it a phase oxide or an adsorbed layer of an oxygen species? In spite of several experimental studies on the subject no clearcut answer was deduced by Gilman in his analysis of the data available at that time [2]. In contrast to Hoare [1], he does not favor the dermasorbed oxygen theory of Schuldiner and Warner [11] which postulates the dissolution of oxygen atoms in the platinum bulk metal.

The anodic oxidation of platinum is believed to occur by the following sequence of elementary electrochemical steps:

$$Pt + H_2O \rightarrow Pt\text{-}OH \tag{2}$$

$$Pt\text{-}OH \rightarrow Pt\text{-}O + H^+ + e \tag{3}$$

$$Pt\text{-}O + H_2O \rightarrow Pt\text{-}O_2 + 2H^+ + 2e \tag{4}$$

In the above mechanism, reaction (2) is assumed to be fast whereas (3) is considered as the rate-determining step (r.d.s.). The appropriate rate equation can be written [2] as

$$i = (1 - \theta_O)nFAk^o \exp [(\beta n_a FE/RT) - m\theta_O] \tag{5}$$

where n is the number of electrons involved in the overall reaction, β the transfer coefficient, n_a the number of electrons in the rate determining step, m the Elovitch or Temkin isotherm constant, and θ_O the fractional oxygen coverage. Here, A is the electrode area, k° the rate constant of the rate-determining step, F and R the Faraday constant and the gas constant respectively, and T the temperature. Neglecting the linear $(1 - \theta_O)$ term, a linear relationship between ℓn i and θ_O arises from Eq. (5), at constant potential. Experimentally, two families of curves are observed yielding two sets of m values. Below 1 V, m equals 38 and above 1.2 V, m equals 15. The change in m value has been related to a change in the stoichiometry of the oxide from PtO_2 [2] in the potential range of 1.0 to 1.2 V.

The relationship between the oxide charge and time (at constant potential), or oxide charge and potential (at constant time) is complex [2]; thus

$$Q_O = \frac{a_s Q_H}{m'} \ell n \left[nFAk^o(t - t_a) \exp \left(\frac{\beta n_a FE}{RT} \right) + \frac{a_s Q_H}{m'} \exp \left(\frac{m'}{a_s Q_H} Q_{Oa} \right) \right] - \ell n \frac{a_s Q_H}{m'} \tag{6}$$

where a_s is the number of equivalents of oxygen per site, $m' = m\, a_s$. Q_{Oa} the minimal value of charge for which the Temkin relationship holds and t_a corresponds to the instant where the Temkin equation is obeyed.

From this equation, the oxide charge is seen to increase very approximately with the logarithm of time. Feldberg et al. [12] did observe a good linear dependence of Q_O with log t for constant E (in the range of 1.25 to 1.65 V) between 1 and 100 sec. This logarithmic growth law will be discussed in greater detail in Section II.B.1.

Another mechanistic approach was proposed by Visscher and Devanathan [7] who considered the oxide growth mechanism to be controlled by the incorporation of metallic cations in the PtO_2 lattice. However, their experimental data (coverage versus potential) were very different from those observed by several other authors [3,5,6]. The electrode coverage corresponding to two layer of PtO (i.e., $Q_O/^SQ_H = 2$) was reached, according to several workers [3,5,6], at about 2 V compared to 1.1 V in the work of Visscher and Devanathan [7]. It was concluded that more definitive experiments were needed to clarify this and other discrepancies in the work on the nature of the oxide film on platinum and the mechanism of its formation.

Since the publication of Gilman's review, a great deal of new work has appeared and many other novel techniques (spectroscopic an optical) have been used to clarify this problem. A survey of these more recent studies as well as the current status of the subject ar given in the following sections.

B. Recent Work (1967-1974)

In the last few years, refined electrochemical techniques as well a new methods of surface analysis have been applied to the anodic oxide film formation on platinum and other noble metals. Among these new techniques are ellipsometry, modulated reflection, ESCA (electron spectroscopy for chemical analysis), LEED (low energy electron diffraction), and others. In the following sections we will discus the new experimental evidence obtained by the foregoing and other techniques in relation to the existing theories on the nature and mechanisms of the oxide film formation on platinum.

1. Electrochemical Measurements

Recently, two groups of investigators have published elaborate expe mental data and discussions on the nature of the oxide film and the mechanism of such film formation and reduction. Vetter and Schultz [13,14] and Schultze [15] have reported some detailed galvanostatic potentiostatic studies on the oxidation of platinum. They favor a place-exchange mechanism coupled with high field ionic transport an they detect a true phase up to three oxide layers thick. It may be added that a place-exchange mechanism was first suggested for the gas-phase oxidation of aluminum [16] and was subsequently applied to the oxidation of several metals at the metal-gas interface [17].

Angerstein-Kozlowska et al. [18] and Tilak et al. [19] studied the potentiodynamic profiles for the same system. The range of potentials covered was limited in most of their work to 1.4 V (versus the hydrogen reversible electrode, RHE). The first stages of the oxidation were carefully monitored under experimental conditions of very high purity. These two important groups of studies [13-15,18,19] are now discussed in detail and in relation to other relevant work on the subject.

In a very pure system [18] the potentiodynamic profile of platinum in acidic medium is illustrated in Figure 1. Several well known features, as also found in other publications [1,2], are evident but a more detailed resolution of the anodic oxide peak is observed. In the region from 0.8 to 1.1 V, three peaks are observed in agreement with the work of Biegler [21]. In this potential range, the oxide coverage was also determined and is shown in Figure 2. The coverage potential curve is very similar to the one obtained previously by Gilroy and Conway [22] from the galvanostatic reduction of the oxide and by Biegler and Woods [23] by a similar technique. The data of

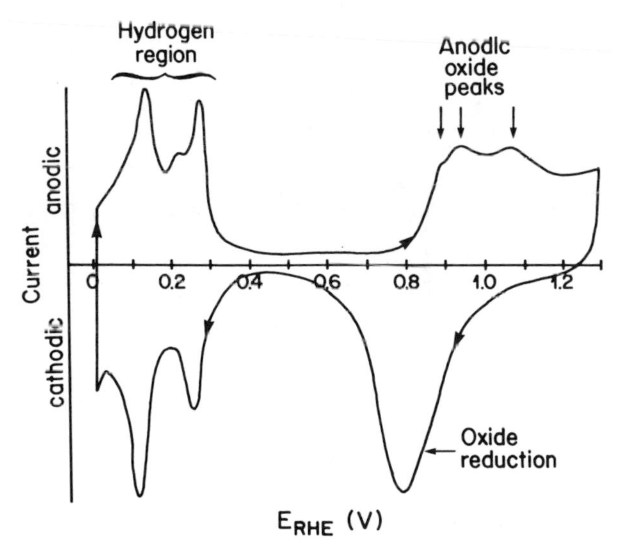

FIG. 1. Cyclicvoltammogram of a bright platinum electrode (after Angerstein-Kozlowska et al. [18]; 0.5 M H_2SO_4 (25°C); sweep rate v = 100 mV sec^{-1}).

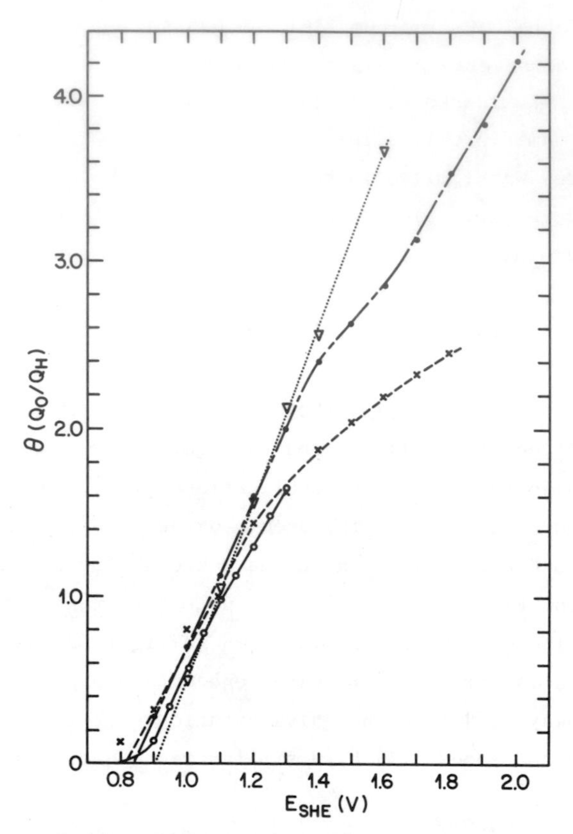

FIG. 2. Oxide coverage of platinum electrodes as a function of electrode potential; o, Angerstein-Kozlowska et al. [18]; x, calculated from the data of Gilroy and Conway [22]; •, Biegler and Woods [23]; and ∇, Vetter and Schultze [13].

Vetter and Schultze [13] are comparable with these [22,23] for oxidation times longer than 10 sec. One criterion for assessing the purity of the system is the equality of charge between the anodic oxide formation up to 1.2 V and the charge needed for its reduction. This equality was not obtained by Vetter and Schultze [13] but they do not specify the upper potential limit of their oxidation. As the anodic potential increases, difficulties arising from oxygen evolution and the anodic dissolution of platinum [24] can be encountered which would lead to an inequality of anodic and cathodic charges.

In the work of Angerstein-Kozlowska et al. [18] the first stages of the anodic oxidation were examined by the cyclicvoltammetric method: the anodic limit of their scanning was generally 1.20 V but in some cases, it was 1.50 V. The oxide coverage, measured by the ratio Q_O/Q_H, versus potential starts in the range of 0.8 to 0.9 V and increases linearly with increasing anodic potential. The charge Q_H is associated with the discharge of protons and is indicative of the available active surface sites. The charge Q_O is associated with the oxide formation. The ratio between the two charges is a measure of the oxide coverage. The values Q_O and Q_H are measured during the same potential scan; Q_O is the integral of the current from the onset of the oxide formation to the limit of the potential scan, and Q_H is the integral of the current between 0.4 V (the onset of the hydrogen ion deposition current) and the onset of the hydrogen evolution reaction. In this work, a break (change of slope) is observed approximately above $\theta = 1$ and at potentials anodic to 1.1 V. Such a break had not been reported by other investigators [13], although Biegler and Woods [23] as well as Gilman [2] did observe such a discontinuity at the much higher coverage of $\theta = 2$.

From different potential scanning programs, it was demonstrated [18] that the first peak in the anodic oxidation can be related to reversible oxide formation. If the anodic scanning is stopped at 0.9 V, a reversible cathodic peak, similar to hydrogen ion deposition and ionization, is observed. If the potential in this type of scanning program is stopped at a higher anodic limit (>0.9 V), the irreversible nature of the oxide formation-reduction is observed, i.e., the cathodic peak for the oxide reduction occurs at a lower anodic potential indicating a higher overpotential for the reduction than for the oxide formation. This irreversibility is observed at an anodic potential of 1.0 V in multisweep experiments. These observations are illustrated in Figure 3.

The three anodic peaks in the oxide formation region of potential (Fig. 1) are attributed to the gradual surface oxidation according to the following mechanism [18]:

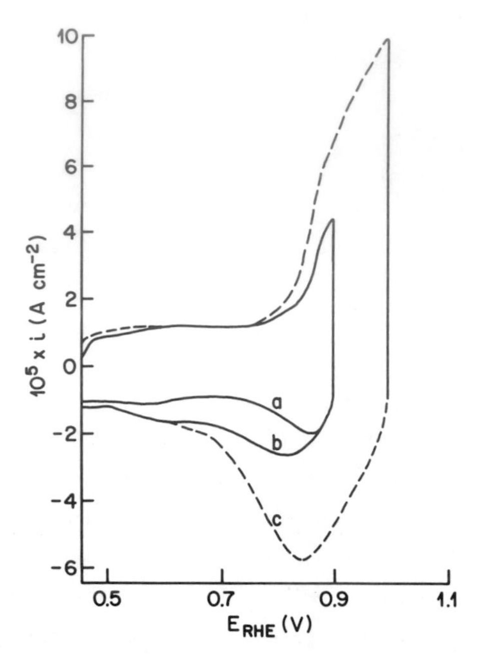

FIG. 3. Potentiodynamic scan of a bright platinum electrode; v = 100 mV sec^{-1}; curves a and c refer to multisweep cycling with anodic limits of 0.89 V and 1.0 V respectively; curve b corresponds to reduction after polarization at constant anodic limit (0.89 V) for 60 sec.

$$4Pt + H_2O \rightarrow Pt_4OH + H^+ + e \quad (7)$$

$$Pt_4OH + H_2O \rightarrow 2Pt_2OH + H^+ + e \quad (8)$$

$$Pt_2OH + H_2O \rightarrow 2PtOH + H^+ + e \quad (9)$$

These species are not necessarily stoichiometric entities but represent surface site occupancy ratios. Associated with these structures, a charge per structure or charge at peak maxima (where the coverage at peak is half the full coverage [25] for a given species) can be theoretically predicted and compared with the experiments. These comparisons are made in Table 1.

A full coverage for a one-to-one correspondence of hydrogen or

TABLE 1

Comparison of Theoretical and Experimental Charges at Anodic Peaks

Structure	Theoretical charge per structure ($\mu C/cm^2$)	Theoretical charge at peaks ($\mu C/cm^2$)	Experimental charges at peaks ($\mu C/cm^2$)
Pt_4OH[a]	55	27.5	26-35
Pt_2OH[b]	55	82.5	81-88
PtOH[c]	110	165	172

[a] The theoretical charge at peak corresponds to half the complete theoretical charge of the Pt_4OH structure.

[b] The theoretical charge at peak includes the total charge for the Pt_4OH structure plus half the charge for the Pt_2OH structure.

[c] The theoretical charge at peak includes the total charge of Pt_4OH and Pt_2OH structures plus half the charge for the PtOH structure.

hydroxyl radical with platinum corresponds to 220 $\mu C\ cm^{-2}$. Thus at the peak maximum where $\theta = 1/2$, the charge associated with the species PtOH will then be 110 $\mu C\ cm^{-2}$. For the structure Pt_2OH, the full coverage by such species would correspond to 110 $\mu C\ cm^{-2}$ so that at the peak maximum the charge is expected to be 55 $\mu C\ cm^{-2}$. The same argument applies to the Pt_4OH structure. It is clear that a close correspondence is observed for the calculated and experimental values (Table 1); other mechanisms which could not generate the three observed peaks were eliminated [18]. The eliminated mechanisms were [18]

$$Pt + H_2O \rightarrow PtOH + H^+ + e \tag{10}$$

followed by

$$PtOH \rightarrow PtO + H^+ + e \tag{11}$$

or

$$PtOH + Pt \rightarrow \begin{matrix} Pt \\ Pt \end{matrix}\!\!>\!O + H^+ + e \tag{12}$$

The above schemes will give rise to charge values much too high compared to the experimental observations and also they cannot generate the characteristic structure (three maxima) in the anodic profile.

The potentials of these peaks are almost completely independent of the sweep rate indicating thereby their reversible nature: these peaks compare well, in this respect, to the hydrogen adsorption-oxidation peaks. In a subsequent study, Tilak et al. [19] performed a detailed kinetic analysis of the above reactions and solved numerically the equations describing the suggested platinum oxidation reactions. By adjusting different parameters (namely the interaction parameter between the adsorbed species PtOH or/and PtO) they could generate the experimentally-observed i versus V curve; their numerically-generated curve, like the experimental curve, was independent of the sweep rate.

As mentioned above, the anodic oxide formation below a monolayer of PtOH is reversible. However, if the monolayer limit is exceeded or if the electrode is polarized at a constant potential by holding the anodic limit of the scan for a definite period of time, the oxide rearranges to a thermodynamically more stable form which gives rise to an irreversible behavior in the reduction transient. The authors [18,19] favor a rearrangement of the PtOH layer into a more stable form by a place-exchange mechanism. This rearrangement will stabilize the species and a higher overvoltage will be needed to reduce this stabilized species. In other words, the isolated PtOH deposition is reversible, but as this species gets surrounded by other PtOH groups, rearrangement occurs and the system acquires irreversibility. This conclusion was not reached by Goldstein et al. [20] from experiments where the oxide formation was limited to 150-1,150 μsec at potentials from 0.85 to 1.1 V. In these experiments, Goldstein et al. could resolve the oxide reduction into two peaks. They interpret these results as a two-step oxidation process where PtOH is formed first, followed by PtO.

The oxide reduction peak was also studied in great detail [18, 19]. The potential of this peak varies linearly with the logarithm of the sweep rate, and the slope of this linear variation was shown by Srinivasan and Gileadi [25] to be related to the steady-state Tafel slope. For the case of the platinum reduction peak, this slope is of the order of 40 mV. This rather low value suggests a more complicated mechanism than the initial discharge mechanism for which a Tafel slope of 120 mV per decade is expected [22]. It is worth noting that a Tafel slope of this order of magnitude (i.e., 40 mV per decade) was also observed by Vetter and Schultze [14].

To account for these observations the following mechanism was postulated [19,22]:

$$\text{"OPt"} + H^+ + e \rightarrow \text{"OHPt"} \tag{13}$$

$$\text{"OHPt"} + H^+ + e \rightarrow Pt + H_2O \tag{14}$$

To generate the experimental curves, reaction (13) has to be in quasiequilibrium and reaction (14) the r.d.s. Only if the standard potential for the "OPt" (~0.4-0.6 V) is less positive than for "OHPt" (~0.8 V), will a single narrow reduction peak be generated [19] with a peak potential variation of 40 mV per decade of sweep rate, as is indeed observed experimentally in Refs. 18 and 22. These species refer to the rearranged electrodeposited entities and differ in energy from the newly-deposited PtOH. If the potential for the OPt reduction was more positive than the reduction potential of OHPt, a double-peak structure would be generated by the solution of rate equations appropriate for reactions (13) and (14).

This behavior (i.e., Tafel slope of 40 mV) is observed only for the reduction of oxide corresponding to two electrons per platinum atom, i.e., for PtO or OPt. For lower coverages or a lower oxidation state, the b (Tafel slope) values range from near-zero to 27 mV [18].

The near-zero b values arise when the surface is oxidized to less than 0.15 e per platinum atom; in this range, the oxidation is reversible [18]. For the range 0.15 to 0.9 e per platinum atom, the Tafel slope tends toward a value of 27 mV in repetitive multisweep experiments. This value increases to 55 mV when the electrode is held at a given potential (>0.9 V) for some time or for a prereduced oxidized surface. For these situations, Tilak et al. [19] offered a semiquantitative explanation. The 27-mV slope is explained by the possible reverse place-exchange mechanism, namely:

$$OHPt \rightarrow PtOH \quad (15)$$

$$PtOH + H^+ + e \rightarrow Pt + H_2O \quad (16)$$

With step (15) in quasiequilibrium with a transfer coefficient of 1, and (16) as the r.d.s., a Tafel slope of 24 mV [$2.3RT/(2 + \beta)F$] is expected, which is quite close to the experimental value.

When reaction (15) becomes the r.d.s. (at faster sweep rates or at more cathodic potentials), a (2.3 RT/F) slope (59 mV) is to be expected; this order of magnitude is observed at higher sweep rates. The above mechanism seems more consistent with the observation that at low temperatures the cathodic peak can be resolved into two separate peaks, one associated with the reversible nature of the surface oxidation and the other, at more cathodic potentials, related to the irreversible nature of the rearranged oxide.

The zero Tafel slopes arise for the reversible component observed when the anodic limit for the platinum oxidation remains near 1 V. On increasing this limit, the second irreversible peak is obtained and the Tafel slope increases.

The main conclusion from this work [18,19] is that for the low oxide coverages, a reversible component of the oxidation could be detected at potentials below 1.1 V where the coverage defined as Q_O/Q_H is lower than 1. However, as the anodic limit of the scanning potential is increased, the reversible PtOH deposited is rearranged by a place-exchange mechanism giving rise to the irreversibility as indicated by the hysteresis in the current-potential profile. A scheme for the anodic oxidation up to PtO (2 e per platinum site) is proposed in which the three anodic peaks observed in the potential range 0.85 to 1.1 V are accounted for. Also, the cathodic reduction of this PtO or the rearranged OPt oxide is proposed where the experimental Tafel slope and shape of the peak (narrow peak) are reproduced theoretically. The coverage (as defined above) versus the electrode potential line is linear with a change of slope at 1.1 V where the coverage is 1. No limiting coverage is observed as in the work of Biegler and Woods [23] but the maximum anodic potential examined was much lower than in the latter investigation.

In another study of the same system, Vetter and Schultze [13-15] reached somewhat different conclusions. They measured the galvanostatic reduction of oxide formed at potentiostatically-controlled different potentials and also studied the potential-time relations for the oxide growth. The aim in their work was different from that of Angerstein-Kozlowska et al. [18] and Tilak et al. [19] in that the former authors polarized platinum to higher anodic potentials since they were interested in the oxide growth and not in the mechanisms of the initial oxide formation. They observed a linear relationship between the oxide charge and electrode potential for a range of polarization times from 2 msec to 1,000 sec. They observed a logarithmic growth law for at least 5.5 decades (2 msec to 1,000 sec) and for potentials from 1.0 to 2.0 V. In these investigations the relationship was linear for the time scale studied, a result different from that of Gilroy and Conway [22] where the linear logarithmic law was observed for polarizations for 1 sec or more. This discrepancy has not been accounted for. The only difference between the foregoing two studies is in the cathodic charging current density used to reduce the oxide. The former authors [13] used low current density and, therefore, longer time for the *oxide reduction* (10-1,000 sec) whereas the latter investigators [22] worked in the millisecond time range for the oxide reduction. This difference in time scale can be important because the longer the measurement period, the more important can be the impurity effects (see Section II.B.2 for such discussion).

An important difference is also observed as the potential increases; it is now well established that the oxide coverage tends to reach a plateau at high anodic potentials (>2 V) [23,26-35].

This observation is not evident from the data of Vetter and Schultze [13] as shown in Figures 2 and 4. It is worth noting that the limiting oxide coverage was observed only when the potentiodynamic sweep method was used for the simultaneous measurement of the oxide reduction and the hydrogen coverage. The galvanostatic method seems less sensitive for this type of measurements. As seen from Figures 2 and 4, the agreement at low (<1.2 V) potentials is very good but at higher potentials discrepancies become important. One problem related to these measurements is the value ${}^{S}Q_H$ (i.e., the saturation coverage by hydrogen where one hydrogen atom is associated with one active platinum atom). The exact determination of these data is of the utmost importance. Using the data of Gilroy and Conway [22], two types of coverage-potential curves can be detected as shown in Figure 5. For a constant value of ${}^{S}Q_H$, the coverage-potential

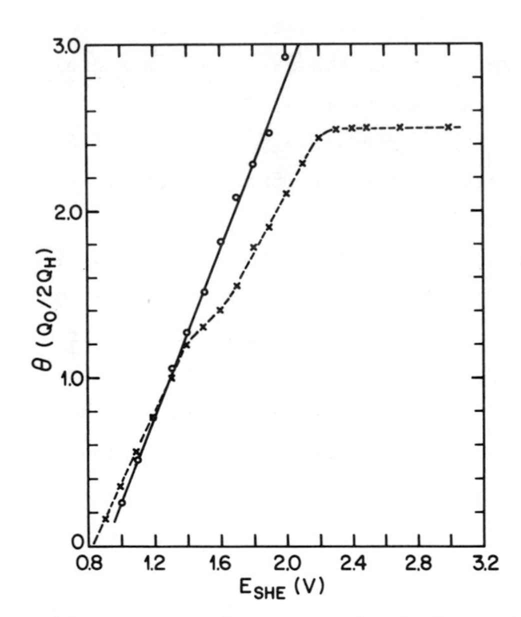

FIG. 4. Oxide coverage for a smooth platinum electrode (25°) at a high (>1.5 V) anodic polarization; o (by galvanostatic reduction) after Vetter and Schultze [13]; and x (by potentiodynamic reduction) after Biegler and Woods [23]. Time of anodic polarization is 10 sec in both cases.

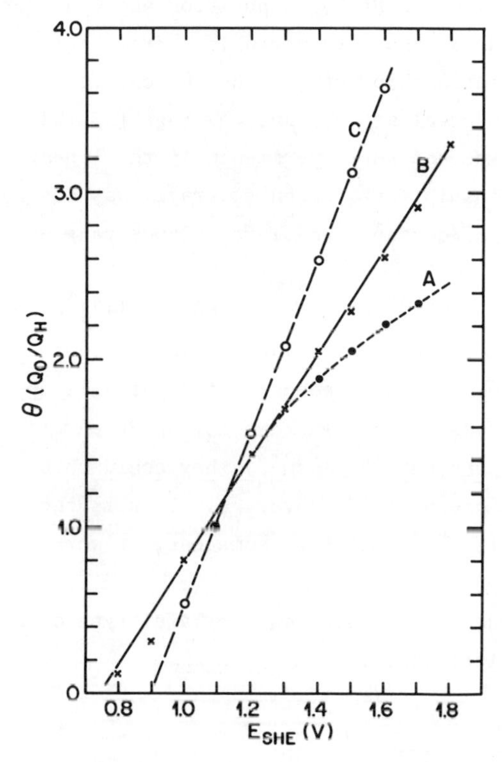

FIG. 5. Oxide coverage discrepancies among different authors: A and B, Gilroy and Conway [22]; C, Vetter and Schultze [13]; B and C for constant value of ${}^{S}Q_H$, A for variable ${}^{S}Q_H$ as determined experimentally [22]; anodic polarization time was 10 sec.

relationship is linear and does not indicate any break or change of slope. However, an increase in ${}^{S}Q_H$ was observed as the potential was increased and a break in the curve was, therefore, noted in the potential range of 1.2 to 1.6 V; this type of behavior has also been observed by other authors [2,23]. The increase in the ${}^{S}Q_H$ value can be traced back to incomplete reduction of the oxide or, most likely, to the presence of a second type of oxide that is more difficult to reduce; this will be discussed in greater detail shortly.

From the linear variation of coverage with potential at constant-time formation and with the logarithm of time at constant-potential formation, Vetter and Schultze [13] concluded the presence of only a phase oxide of $Pt^{II}O$ composition and considered the thickness of this oxide as the main variable. From their coverage data, they deduce a relationship between the current for its formation and the electrode potential at constant coverage [13,15]. A Tafel type of relation is expected but experimentally the linear log i versus potential relationship (at a given coverage) was observed only over a limited number of decades, the Tafel slopes ranging from 40 to 140 mV.

In a further study, Vetter and Schultze [14] examined the oxide formation and reduction by a galvanostatic procedure in which a galvanostatic step of the required current density is applied to an oxide formation constant current density of 20 μA cm^{-2}. By such procedure (at a constant 20 μA cm^{-2}) they could obtain the oxide overvoltage at a preset oxide coverage. This is the same procedure as used by Schultze and Vetter in some earlier work on the anodic oxide formation on gold [36].

From these data they observed the Tafel type of behavior and the following relationship:

$$\log i_+ = A_+ + \frac{\varepsilon - E_+}{b_+^o(1 + a_+\theta)} \tag{17}$$

where i_+ is the applied oxide formation current density, ε is the measured potential, E_+ is the oxide formation potential of 0.53 V, b_+^0 and a_+ are constants (36 mV and 1.0 respectively), $A_+ = -16$, and θ is the coverage. Experimentally, the Tafel slope obeyed the following law:

$$b_+ = b_+^o(1 + a_+\theta) \tag{18}$$

The same type of relation was also obtained from the potentiostatically-determined curves [13]. The A_+ and E_+ are obtained by the extrapolation of each Tafel line for different coverages and the convergence point occurs at $\log i_+ = -16$ and at 0.53 V.

The cathodic (oxide reduction) experiments yielded different results [14], namely, a different Tafel slope behavior; a Tafel slope of 60 mV independent of coverage for $\theta > 1$ was observed.

The relation between the current density i and coverage at constant anodic potential did not yield the theoretically predicted relationship

$$\log i = \log i\,(0) - k\theta \tag{19}$$

the experimental curves (log i versus θ) for the anodic oxide formation being slightly curved. This curvature was attributed to time effects (aging, reorientation of the oxide layer) in the oxide formation and reduction [14]. For the cathodic reduction of the oxide the relationship between the applied current density and coverage was of the form

$$\log |i_-| = \log |i_-(0)| + k_1\theta \tag{20}$$

Such a relationship had been previously observed by Bagotzki et al. [37] for the reduction of platinum oxide.

From the above data and other observations Vetter and Schultze [14] concluded that the oxide on the electrode was a phase oxide. Some additional facts leading them to this conclusion were (i) the corrosion of platinum under polarization [38] indicated the presence of movable platinum ions; (ii) the roughening of the surface and corrosion during *reduction* of the oxide layer [39,40] are indications of a phase oxide, according to these authors [14]; (iii) the ellipsometric evidence [41] was cited in favor of the phase oxide theory since a phase oxide should have different reflection properties from those of a bare metal or a chemisorbed oxide layer.

Vetter and Schultze [14,15] used a particular double layer model and the equivalent electrical circuit to illustrate their assumption of the formation of a phase oxide of constant composition with the oxide thickness as the main variable. The model and the circuit are shown in Figure 6 in which C_{ad} and R_{ad} refer to the adsorption of oxide ions in the following equilibrium reaction:

$$H_2O(aq) \rightleftarrows O^{2-}(ad) + 2H^+(aq) \tag{21}$$

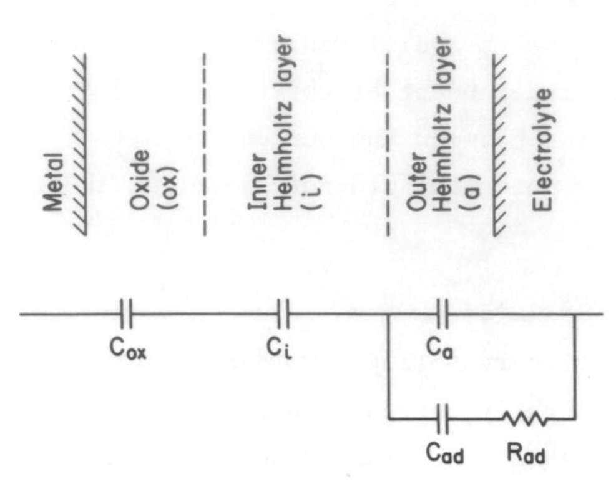

FIG. 6. Equivalent circuit representation for high-field ionic conduction model [14].

In the absence of oxide formation, the double layer capacitance C_D is given by $1/C_D = 1/C_i + 1/C_a$ where C_i and C_a are the capacitances of the inner and outer Helmholtz layer. As the potential increases, the oxide will be formed and C_{ox} has also to be considered. The value of C_{ox} decreases with increasing thickness (d) of the oxide layer in such a manner that $C_{ox} = D_{ox}/4\pi\delta_{ox}$ where $\delta_{ox} = \theta d$ and D_{ox} is the dielectric constant of the oxide film. At this stage then

$$\frac{1}{C_D} = \frac{1}{C_i} + \frac{1}{C_{ox}} = \frac{1}{C_i}(1+ a\theta) \tag{22}$$

Such a relationship between $1/C_D$ and θ (where a is an experimental constant) was indeed observed by several workers [14]. In Eq. (22), the C_a is considered to have been short-circuited by the equilibrium reaction (21). From this model a coverage of the oxide ion (O^{2-}) was calculated to be 0.1 at 1.5 V.

In this type of description, the r.d.s. is the place-exchange between a platinum ion at the metal surface and the oxygen ion in the inner Helmholtz layer for $\theta \ll 1$. As the surface reaches full PtO coverage, the oxide growth continues by a similar mechanism with the r.d.s.'s being postulated at the metal-metal oxide interface or at the metal oxide-inner Helmholtz layer interface. By this type of mechanism a continuous oxide growth, indicated by the absence of any limiting oxide coverage, is accounted for [13-15]. Such field-assisted oxide growth was also postulated by Visscher and Devanathan [7] and Ord and Ho [42]. A similar interpretation was used by Inman and Weaver [43] to explain the oxide growth on platinum in LiCl-KCl at 425°C with LiO_2 as the electrolyte. The anodic oxide formation is accounted for by a reversible oxide deposition reaction followed by the rate-determining rearrangement of the oxide constituents by a place-exchange mechanism, without involving any electron transfer. This model of anodic oxide growth was also used by Damjanovic et al. [44] and Harris and Damjanovic [45] in their study of platinum oxide growth in oxygen-saturated acid solutions.

Inman and Weaver [43] also observed an ageing effect indicating the presence of a rearrangement and the typical hysteresis between formation and reduction was also noted. The strongest evidence for the above type of mechanism comes from the nondependence of the initial oxide formation on potential, indicating a reversible electrochemical mechanism. However, this reversible reaction has not been clearly identified although it is believed to involve the discharge of the O^{2-} ion. Also the interfacial capacitance values (as measured from the variation of potential with time on open-circuit decay) are much higher than those expected for double-layer charging process only. An adsorption pseudocapacitance component is thus likely to be involved.

For the oxide reduction in this melt, Inman and Weaver [43] could not propose a simple reaction mechanism. To obtain a good fit to a simple desorption rate equation, the authors had to invoke unrealistic reaction orders [5 to 6]. With more reasonable values of reaction orders, the heterogeneity factor had to be taken as nonnegligible, as was shown by Stonehart et al. [46]. Some studies on the oxide formation-reduction mechanisms [47,48] at high temperatures have also been reported. Calandra et al. [47] and De Tacconi et al. [48-50] used $KHSO_4$ and $NaHSO_4$ melts in the range of 180 to

300°C. Their results are best explained by the presence of a water formation equilibrium of the type

$$2KHSO_4 \rightleftarrows H_2O + K_2S_2O_7 \tag{23}$$

The water released in reaction (23) is the reacting species which produces the oxide on platinum. Calandra et al. [47] and De Tacconi et al. [48] observed behavior similar to that for aqueous systems, e.g., the hysteresis between the oxide formation and reduction and ageing effects. At high temperatures they also observed some chemical dissolution of the oxide during its electrochemical reduction. For the oxide reduction mechanism, De Tacconi et al. [49,50] proposed a rapid electron transfer reaction followed by a second order recombination step under Temkin conditions.

As mentioned previously, the model proposed by Vetter and Schultze [13-15] is not likely to be universally accepted. One important experimental observation that was overlooked by these authors was the limiting oxide coverage observed by several workers [23,26-35]. In a careful critical study, Biegler et al. [51] found a limiting oxide coverage at potentials greater than 2 V, corresponding to a coverage $Q_O / 2^S Q_H = 2.15$ (i.e., two oxygen atoms per platinum surface atom), and they concluded that this was due to a monolayer of PtO_2 oxide. This result was obtained by a critical assessment of the hydrogen limiting coverage, the crucial importance of which was mentioned earlier.

A limiting coverage of 2.08 is also observed for platinized platinum [51]. From this result Biegler et al. [51] concluded that the formation of a chemisorbed monolayer of oxygen film with the composition PtO_2 was involved since, they argued, no limiting coverage could be accounted for if phase oxide formation were assumed. For a phase oxide mechanism the film should grow to a thickness greater than a layer or two. They do not provide a mechanism for the formation of such a limiting coverage, however. This limiting coverage was studied as a function of the sulfuric acid concentration by Visscher and Blijlevens [34]. They observed a decrease in the oxygen coverage as the sulfuric acid concentration was increased from 1 M to 11 M, this decrease being linear with increasing acid concentration but not with respect to the HSO_4^- concentration. The limiting oxygen coverage was observed at about 2.1 V (versus RHE). This result indicates the competition between the acid or sulfate adsorption on the one hand and the oxygen on the other, and the limiting coverage seemed to correspond to a passivating layer with the composition $PtO_x(H_2SO_4)_y$ where x and y depend on the acid or sulfate concentration.

Momot and Bronoël [52,53] claimed to have detected the presence of a platinum oxide PtOH in the range of 0.4 to 0.8 V (versus the standard hydrogen electrode, SHE) in acidic medium employing a potential step method where, subsequent to a potential sequence to clean the electrode, the potential was held at 0.375 V and a potential step to the required value was applied. The current associated with the potential step is integrated between the time when the potential step was applied to the time when the current reached a steady state, and a coverage versus potential relation can thus be obtained. However, these results did not agree with the earlier work of Icenhower et al. [54] using an identical method. The cause of this discrepancy could be the presence and oxidation of organic impurities, and especially undercompensation of the double-layer charging in the experiments of Momot and Bronoël [52,53]. These discrepancies cast doubts as to the validity of the interpretations of Momot and Bronoël.

Another intriguing experimental fact is observed for electrodes polarized to high anodic potentials for long times; under these conditions an oxide reduction peak appears in the potential region of the hydrogen ion deposition, namely, in the potential range of 0.03 to 0.4 V. This type of oxide has been observed by several workers [29-32,37,55-61]. The potential program and the resultant voltammetric curve is illustrated in Figure 7. This type of oxide, denoted as type II by Balej and Spalek [32], is not obtained just by high anodic polarization; a mechanical stress (induced by polishing

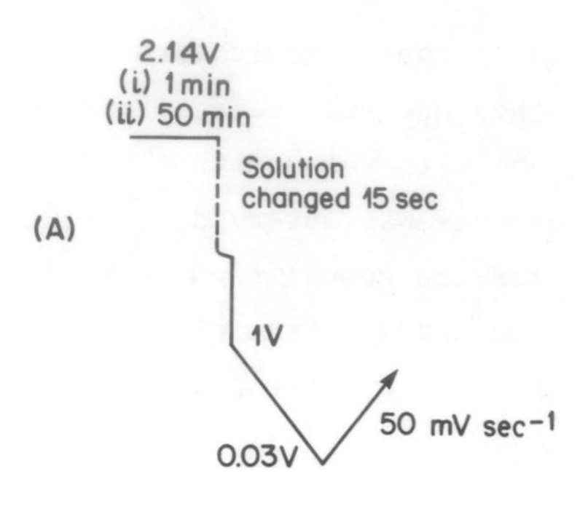

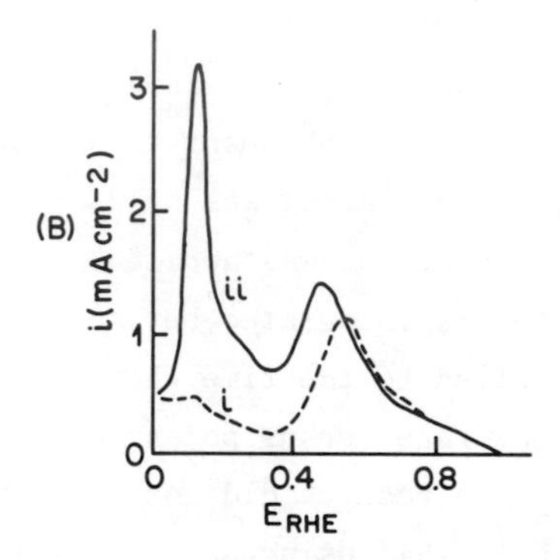

FIG. 7. (A) Program and (B) cyclicvoltammogram illustrating the β or type II oxide formation [32]; 0.5 M H_2SO_4, ground platinum electrode. (i) 1 min and (ii) 50 min anodic polarization at 2.14 V.

or grinding) in the electrode surface seems to favor its formation [32]. Depending on the experimental conditions (high anodization potential and current and long times of polarization, i.e., several hours) two peaks corresponding to the oxide phase reduction appear at 0.2 V and 0.3 V [59]. The detailed analysis of this type II oxide was performed by Balej and Spalek [32] who showed that the oxide grows linearly with the anodization time. Also, the initial rate of growth varies with potential; a maximum is observed at 2.3 V in 2.5 M H_2SO_4. At potentials lower or higher than 2.3 V the growth rate is diminished. As stated above, the formation of this oxide is very much dependent on the pretreatment of the electrode; a platinum electrode preheated in an oxidizing flame to a yellow glow will not sustain the formation of this type of oxide even on polarization at 2.2-2.4 V for several hours. However, grinding the electrodes will induce the oxide II formation easily.

The oxide II formation obeys a parabolic growth law of the form [32]

$$(Q_O^{II})^2 + k_1 Q_O^{II} = k_2 \tau \qquad (24)$$

where Q_O^{II} is the amount of oxide deposited as measured by its reduction charge, τ is the time of polarization, and k_1 and k_2 are constants. This parabolic law is derived for cases where the r.d.s.'s are both diffusion of ions through the oxide layer and the chemical formation of the oxide [32].

The chemical identity of the oxide is still open to speculation. The ratio $Q_O / 2^S Q_H$ does not reach a limit and can be as high as 20 for a polarization time of 17 hr at 2.14 V. It seems, therefore, that one has really a phase oxide in this case as compared to a monolayer of chemisorbed oxygen for the first oxide reduced at 0.6 V. Electron diffraction experiments by Shibata [62] have indicated the presence of a PtO_2 oxide. However this result has never been confirmed electrochemically as the electrochemical behavior of the chemically-produced PtO_2 did not reproduce the oxide II type of behavior [32].

It may be pointed out that a similar set of data was obtained by Shibata and Sumino [60] when a maximum in anodic charge associated with the type II oxide (or β oxide as they called it) was observed. The maximum was observed at 2.14 V (versus SHE) compared to 2.3 V as reported elsewhere [32]. This discrepancy can be explained by the difference in the sulfate concentration of the solution, and, most importantly, by the electrode pretreatment. Shibata and Sumino [60] annealed their electrodes in vacuum at 1,400°C for 1 hr, whereas in the work of Balej and Spalek [32] the platinum electrode had to be ground before the type II oxide could be observed on anodization. The former authors [60] interpret their data as a passivation phenomenon. That the temperature of the electrode polarization plays an important part was shown by several workers [60,61]. Visscher and Blijlevens [61] reported that for electrodes that were not oxidized previously, the temperature at which the oxide II could be detected for polarization at 100 mA cm^{-2} was around 15°C. For electrodes that were preoxidized anodically, the temperature had to be at least 22°C for the oxide II formation to be observed on anodization at high anodic potentials. The oxide II was also observed on platinum under cathodic polarization in alkali nitrate melts by Shibata and Sumino [63].

A very important concept in the problem of anodic oxide formation on platinum is the possibility of absorbed oxygen within platinum, the so-called dermasorbed oxygen [1,2]. Schuldiner and Warner [11,64] observed two types of absorbed oxygen: the first type is mainly a monolayer on the surface, whereas the second type is what they called dermasorbed oxygen, i.e., oxygen atoms dissolved in the first two or three layers of the platinum lattice. The question of absorbed oxygen was reexamined recently [65,66] using galvanostatic reduction transients on platinum electrodes subjected to different pretreatments. The electrodes were cleaned in a hydrogen flame followed by quenching in concentrated nitric acid. The type of galvanostatic reduction curves obtained by Thacker and Hoare [65] are different from those of Vetter and Schultze [13]. At a potentiostatic polarization of 2.05 V, Thacker and Hoare [65,66] observed a second arrest between 0.3 and 0.16 V, attributed to absorbed oxygen atoms dissolved in the platinum lattice (Fig. 8). However, this arrest could be related to the type II discussed above, formed at a very high current density (~10 A cm^{-2}) corresponding to the high anodic polarization potentials (2.3-3.0 V) reported in the previous studies [29-32,51,55-62].

The fact that Thacker and Hoare [65] could observe this arrest under less drastic anodization conditions can perhaps be attributed to the pretreatment of their working electrode (the hydrogen flame and, especially, quenching in concentrated nitric acid). In their

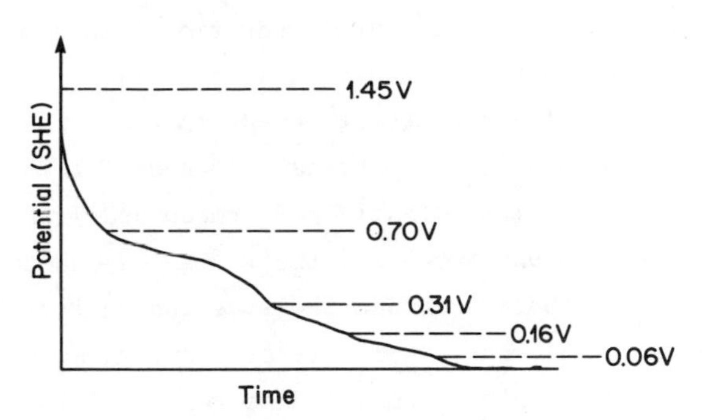

FIG. 8. Galvanostatic oxide reduction curve, after Thacker and Hoare [65]. Anodic polarization at 1.45 V, reduction current density 310 μA cm^{-2}. Identity of potential regions: 0.75-0.31 V, reduction of surface adsorbed oxygen; 0.31-0.16 V, reduction of "dermasorbed" oxygen; 0.16-0.06 V, hydrogen ion adsorption.

data, they observed the formation of a monolayer of PtO_2 ($\theta = 2$) as the electrode is polarized galvanostatically at 62.6 mA cm^{-2} for several hours, as was also reported by Biegler and Woods [23]. However, the stability of the PtO_2 oxide is rather low; at open circuit it will tend to decompose into a more stable form, PtO, with the absorption of excess oxygen into the bulk metal up to a few layers. Further work [66] has shown the stability of the absorbed oxygen as rather high in that it took extended periods (∿2 days) of cathodic treatment (-100 mA cm^{-2}) to remove all the absorbed oxygen. Alternatively, heating to white heat could also be used to reestablish the original behavior of an untreated platinum electrode. The oxide coverage plays an important role in the attainment of reversible potential for the oxygen electrode [1,65-69]. However, the role of platinum dissolution is also considered to be an important factor for obtaining the reversible oxygen electrode potential on a platinum electrode [70].

The role of oxygen dissolved in the platinum lattice with respect to the oxygen gas reduction was studied by Hoare [68]: in this study, the back side of a platinum disk was polarized anodically. After prolonged anodization, the front side of the diaphragm showed an enhancement in the oxygen reduction. This effect was attributed to the new platinum-oxygen alloy [68].

Another important parameter relevant to the study of anodic oxide formation on platinum is the capacitance of this electrode as a function of the applied potential. Formaro and Trasatti [71] made a comparative study involving their results and those of Gilman [2]. The former authors examined the time and purity effects on the capacitance-potential curves. The results are shown in Figure 9. We note the agreement of the capacitance values at 400 mV where the electrode is free of hydrogen or oxygen adsorption. The 34 μF cm^{-2} found is higher than the 18 μF cm^{-2} reported elsewhere [72 and references therein]. The impurities present in the solution seem to be responsible for this deviation. Carr and Hampson [73] observed a value of 20 μF cm^{-2} but their area pertains to the geometric and not the real surface so that a direct comparison of the results is not possible. As the anodic potential increases, the capacitance passes through a maximum at 1.1 V, and with a further increase in potential to 1.5 V, the capacitance value decreases. The increase

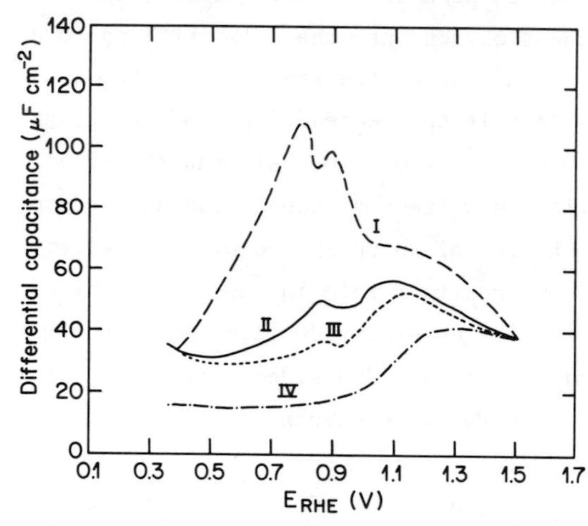

FIG. 9. Differential capacitance versus potential curves for smooth platinum under different experimental conditions (measured after 1 sec, 30 sec, 3 min, and under steady state conditions, respectively). Curve I is due to Gilman [2] and curves II-IV are taken from Formaro and Trasatti [71].

in capacitance in the potential region of 0.7 to 1.1 V is associated with the oxygen adsorption process

$$Pt + H_2O \rightarrow PtOH + H^+ + e \quad (1)$$

Above 1.1 V, the oxide PtO is formed bringing a decrease in the capacitance value. The capacitance value below 0.4 V increases illustrating the hydrogen adsorption [73]. Formaro and Trasatti [71] have shown the important effect of impurities between 0.4 and 0.8 V. The measured differential capacitance decreases with time, the decrease being more rapid in impure solutions for potentials below 0.8 V. The variation of the capacitance with time in the potential range of 0.8 to 1.2 V is very rapid and important even in purified solution. At 0.8 V, the differential capacitance drops from 47 $\mu F\ cm^{-2}$ to 18 $\mu F\ cm^{-2}$ in 10 min. At 1.3 V however, the drop is only 5 $\mu F\ cm^{-2}$ from 48 to 43 $\mu F\ cm^{-2}$. This time effect is believed to be associated with the rearrangement of the surface oxides [71].

In a study of the kinetics of the oxide formation and dissolution on platinum, Ohashi et al. [74] derived a rate equation for the oxide reduction which is of the type

$$i_c = nFk_{11}q^2 \exp \frac{\alpha nFvt}{RT} \quad (25)$$

where i_c is the oxide reduction current during the potential scan at a rate of v (V sec^{-1}), q is the amount of oxide deposited, k_{11} is the second order chemical rate constant, α is the transfer + coefficient for the cathodic reduction of the oxide, and the other symbols are the standard electrochemical constants. One important conclusion arising from Eq. (25) is the prediction of a second order reaction with respect to the oxide coverage; such prediction has been confirmed experimentally [74]. To account for this behavior, Ohashi et al. [74] postulated the presence of hydrogen peroxide as a reaction intermediate in the following sequence:

$$2PtO + 2H^+ + 2e \rightarrow 2Pt + H_2O_2 \quad (26)$$

$$Pt + H_2O_2 \rightarrow PtO + H_2O \quad (27)$$

However, the presence of the peroxide is controversial; some authors were unable to detect its presence [75,76] whereas others did observe its formation [77,78]. The peroxide formation occurs predominantly in the oxygen gas reduction but not necessarily in the oxide reduction process in the absence of oxygen.

Sasaki and Ohashi [81] confirmed the second order rate observation by studying the transient current of the oxide reduction where the expected relationship between the charge and time is of the form:

$$\frac{1}{q} = At + \frac{1}{q_0} \tag{28}$$

where A is equal to k exp $\alpha nFE/RT$. Figure 10 shows the experimental points as well as the variation of the slope A with the potential of reduction. Equation (28) is seen to be followed within experimental error. The transfer coefficient α thus obtained is equal to 0.25 and agrees with the previous results [62]. However, Ioi and Sasaki [79] seem to have neglected the double-layer charging process and this would lead to somewhat different results since they studied the rapid (in the millisecond range) phenomena after a potential step. The single rate equation [Eq. (25)] for the oxide reduction seems too simple to account for the complex experimental evidence observed under different sets of conditions [19,80]. Upon solving Eq. (25) a linear relationship between the current maximum for the oxide reduction and the sweep rate is obtained [74,81]. However, this behavior is very much dependent on the anodic formation of the oxide as shown by Angerstein-Kozlowska et al. [18]; the linearity is observed only when the two cathodic peaks are separated at low temperatures, indicating the complex nature of the reduction process. At room temperatures the oxide reduction peak is only approximately linear with the sweep rate [18].

Most of the work presented so far was carried out on smooth or bright platinum electrodes. In practical use (e.g., fuel cells and electrocatalysis in general) high surface area is sought. The general features for platinum black or platinum deposited on an inert

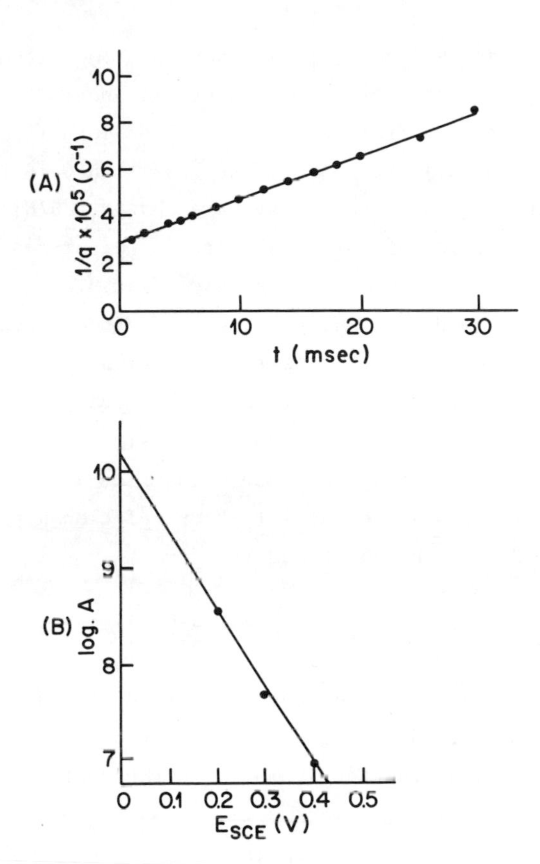

FIG. 10. (A) Inverse of the oxide reduction charge as a function of time for a potentiostatic oxide reduction transient: $1/q = At + 1/q_0$ [79]. (B) Variation of log A as a function of the magnitude of the reduction potential of the transient: $\log A = \log k + \alpha nFE/2.3\ RT$ [79].

substrate are similar to the ones observed on bright platinum. However, only a few studies have been performed on these materials to examine the oxide formation or reduction kinetics. Lundquist and Stonehart [82] have recently studied the oxide formation on different platinum crystallites deposited on graphite. The diameter of the crystallites studied ranged from 30 Å to 100 Å. The voltammetric curves registered are identical with those for the bulk (wire or foil) platinum [83].

Lundquist and Stonehart [82] observed a logarithmic oxide growth at constant potential for potentials above 0.98 V (versus RHE). For potentials from 0.83 to 0.93 V, a logarithmic law is also observed but a limiting coverage less than 0.15 is obtained. From these curves, the rate of coverage change, $d\theta/d(\log t)$, can be plotted as a function of potential; at low potentials (0.8-1.0 V), the different forms of platinum, namely, flat platinum sheet, unsupported platinum black, and platinum on carbon (loading = 72 m^2 g^{-1}), fall on the same line. As the potential increases (or the oxide coverage increases) a distinction between these types of catalysts is observed. The maximum rate is observed for the bulk platinum and the rate decreases as the surface area of platinum crystallites increases. A Tafel type of relationship can be deduced for oxide growth and it has the form

$$\frac{d\theta}{d(\log t)} = k(1 - h\theta) \exp - \left[\frac{(1 - \alpha)nF}{\nu RT}\right]\eta \qquad (29)$$

where k (in sec^{-1}) is the heterogeneous rate constant, h is a composite constant including the heterogeneity parameter, n is the number of electrons transfered, and ν is the stoichiometric number. Plots of $\log [d\theta/d(\log t)]$ versus the formation potential η yield straight lines for potentials in the range of 0.8 to 1.1 V with a slope of $(1 - \alpha)n/\nu = 0.25$.

The oxide reduction peak for various constant oxide coverages was studied as a function of the cathodic sweep rate. The relationships $(dV/dt)t_{i,max}$ versus the log of potential scan rate (log dV/dt) yield straight lines, but with slopes changing with the coverage. At low coverages ($\theta < 0.5$) the slope of such curves gives a Tafel parameter of 30 mV, and as the coverage increases to 1.5, the Tafel slope increases to 50 mV [82]. This change is interpreted as a change in mechanism [82]. Such variation in the Tafel slope for the oxide reduction has also been reported by others [14,18]. The variation in Tafel slope was interpreted [18] as the change from a partly reversible process at low oxide coverage to an irreversible process due to rearrangement of the oxide. Since this behavior is observed for the three types of electrodes, Lundquist and Stonehart [82] concluded that there was no effect arising from the carbon support. However, an effect associated with the crystallite size was noted; a linear relationship exists between the oxide coverage for a given Tafel slope and the percentage of surface atoms per crystallite surface. As the percentage of surface atoms increases (i.e., the crystallites become smaller) the coverage decreases. This behavior reflects a change in the free energy of the surface [82]. It is argued [82] that as the size of crystallites decreases, the platinum surface looses its metallic character and becomes a poorer catalyst. This behavior is confirmed by Blurton et al. [84] in their study of dispersed platinum electrodes where the crystallite diameter was of the order of 14 Å. For such dispersed electrodes the platinum activity for the oxygen reduction is lower than that for a platinum black electrode. The cyclicvoltammetric curve indicates a strong influence of the graphite substrate; the characteristic hydrogen peaks from 0.0 to 0.4 V are not well defined and strong background currents for the oxidation and reduction of active groups on the graphite are observed. The results are shown in Figure 11 where the qualitative similarity between the carbon electrode and the dispersed platinum electrode is evident. Both the platinum black and the platinized platinum behave like bright platinum (Fig. 11).

The limiting coverage by the oxide indicated on bright platinum was also observed for platinized platinum [51] where the limiting value of two (2.08) oxygen atoms per platinum surface atom was found. It can thus be concluded that the increase in the platinum electrode area does not change the mechanism of formation and reduction of oxides. However, at very high values of platinum dispersion on carbon, the platinum loses its metallic character and the electrode approaches the behavior of the inert support; the lowest platinum crystallite diameter for such behavior to be observed is of the order of 14 Å.

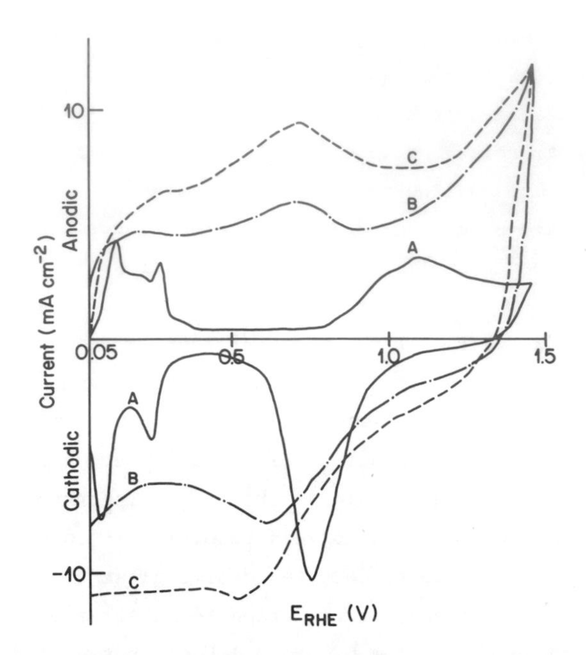

FIG. 11. Cyclivoltammogram of dispersed platinum electrodes in 1 M H_2SO_4 at 30°C, after Blurton et al. [84]. A, Pt black electrode (9 mg Pt per cm^2); B, dispersed platinum on carbon (0.6 mg per cm^2); C, pure carbon electrodes.

2. Optical and Spectroscopic Methods

In recent years, electrochemical studies of the oxide formation and reduction on noble metals and on platinum in particular have also been supported by some complementary nonelectrochemical techniques. The earliest optical technique applied to this problem was ellipsometry. Other optical methods such as electrochemical reflectance, Auger, X-ray, LEED, and ESCA spectroscopy have also been used to obtain information on the formation of oxides, their nature, and the mechanism of their formation or reduction. In the following subsections we will deal with each technique separately.

a. Ellipsometry. The theory of ellipsometry and the principles involved in its application to electrochemical problems have been recently reviewed by Conway [85] and will not be discussed here. The first data on the application of this technique to the oxide formation on platinum were published by Reddy et al. [86] who calculated from ellipsometric results the thickness of the oxide film as a function of potential; the results are shown in Figure 12. From ellipsometric data we note the formation of an oxide with optical properties different from those of the bare metal at 0.98 ± 0.01 V [86]. The coulometric results agree in general with the ellipsometric ones except for the formation potential of the oxide. To calculate the thickness of the film from coulometric or ellipsometric data, assumptions as to the nature of the film (its density, index of refraction, etc.) have to be made, and this can cause discrepancies in the results. The ellipsometric technique has been improved since then and automatic and rapid methods are now used to gather the data [42]. Since several assumptions have to be made to calculate the oxide thickness, many workers publish the actual ellipsometric data and correlate it with the coulometric charge associated with the formation of oxide at a given potential. In the more extended paper by Reddy et al. [87] such measurements were reported and a more elaborate discussion of the mechanism and the

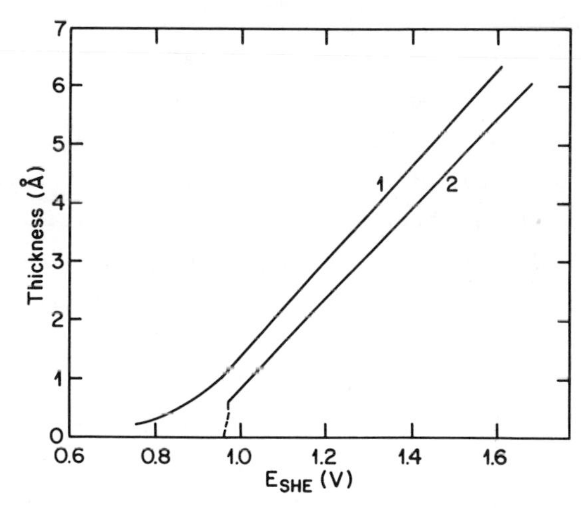

FIG. 12. Oxide thickness as measured by coulometry (1) and ellipsometry (2) as a function of electrode polarization [86].

nature of the oxide formation was presented. To fit their experimental data, a value of 2.625 for the refractive index of the platinum oxide, PtO, was chosen. This is to be compared with values ranging from 2.60 to 2.65 obtained for bulk $PtO.H_2O$ films [87]. We may note that in the experimental data [87], the ellipsometric parameters Δ and ψ remain constant for polarization from 0.4 to 0.9 V (versus SHE). As the potential increases above 0.9 V a linear decrease of these parameters with potential is observed. However, there has been a controversy [88,89] as to the constancy of the optical parameters at electrode potentials from 0.7 to 0.9 V; in this region the onset of the oxide formation occurs and hence its detailed examination is of the utmost importance.

There are several reports on the variation of the relative phase retardation Δ in the potential range from 0.61 to 0.9 V (versus SHE) [18,42,90-95]. However, some other workers did not observe such variation in this low potential range but only a linear variation commencing at a potential of about 0.8 or 0.9 V [86,87,96]. These discrepancies can be attributed either to electrode preparation or to the differences in the sensitivity of the ellipsometric instrumentations. There are two main types of data acquisition procedures in the ellipsometric experiments: the slow (steady-state) and the rapid (nonsteady-state) methods. In the steady-state or slow method, the potential is scanned point by point and the optical data are obtained after the steady-state condition has been reached; the variation of Δ with potential is observed only at high potential. If the optical data are obtained rapidly, the potential is scanned continuously, and the ellipsometric parameter is obtained concurrently, then under these conditions, the variation in the optical parameters is observed at lower potentials. The optical and coulometric results above 1.1 V are in perfect agreement and it is now relatively well accepted that in this potential range the data can be related to the formation of a compound with the stoichiometry PtO. The index of refraction measured for the bulk oxide [96] agrees well with the one obtained from ellipsometric data [86,87,95]. In the potential range of 0.6 to 1.1 V, the interpretations are somewhat different. From their data, Angerstein-Kozlowska et al. [18] argue that the relation between Δ and the oxide coverage illustrates the formation of the PtOH oxide; they find no abrupt change in this relationship but only a minor change of slope at $\theta = 0.5$ and $\theta = 1$ ($\theta = Q_O/Q_H$). For Kim et al. [95], Vinnikov et al. [91], and Chiu et al. [98], variations in Δ indicate anion adsorption preceding the formation of a phase oxide; the potential for the onset of this phase formation is established at 0.98 V.

Figure 13 illustrates some typical ellipsometric and coulometric results obtained [88] in the potential range from 0.1 to 1.4 V at a scan rate of 280 mV sec^{-1}. The data represent an average for four successive sweeps. We note a decrease in Δ from 0.5 to 0.9 V, followed by a greater rate of change at 0.9 V, with still

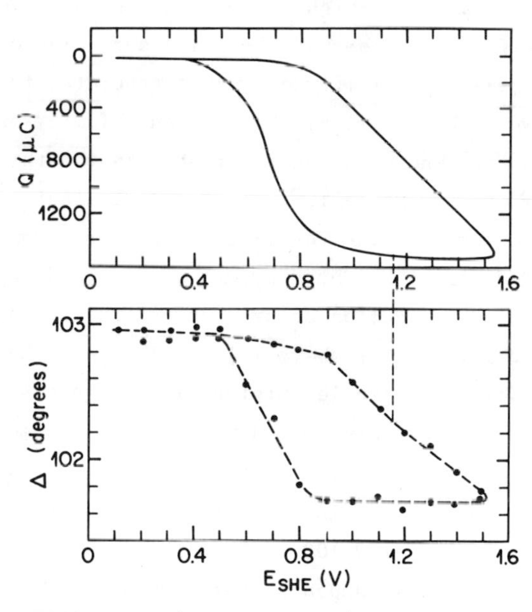

FIG. 13. Ellipsometric parameter Δ (relative phase retardation) and oxide formation and reduction charges versus potential scanning of a platinum electrode [88]; v = 280 mV sec^{-1}.

another change of slope at 1.15 V. These observations were confirmed by Angerstein-Kozlowska et al. [18]. A discontinuity in the Δ-potential curve at 1.1 V was also observed by Ord and Ho [42] and by Kim et al. [95].

There is no disagreement as to the thickness of the film as a function of potential. The width of the oxide film at 1.4-1.5 V is calculated to be of the order of 5 Å [42,76,87,96,97]. It is interesting at this point to look in more detail into the results of Ord and Ho [42]. They observed a relation between the angles of the analyzer and the polarizer for null readings for different potentials from 0.3 to 1.3 V during galvanostatic charging and discharging curves. Their results are illustrated in Figure 14(A). The main observation is that for the increasing anodic potentials, a line AC with two separate sections is observed, and for the cathodic reduction, a straight line DA with a different slope is observed. From the theoretical treatment for the formation and dissolution of films, a straight line should be observed for given values of n (refractive index of the film) and k (absorption coefficient) [86,94]. The family of curves generated for such a system is shown in Figure 14(B) [87]. Such a family of curves is obtained for a film where the only variable is the thickness. In the results of Ord and Ho [42], it is concluded that such an interpretation is not valid. The inflexion at B (corresponding to 1.1 V) can be accounted for either by a growth of a second layer with a different index of refraction or by a change in the index of refraction as the layer grows above 1.1 V. The fact that the reduction curve DA follows a single straight line suggests that only a monolayer is present at point D and this favors the model in which the index of refraction changes as the layer is formed. The same Δ-ψ curves were obtained in the potentiodynamic scanning experiments of Horkans et al. [99,100]. From the results of Angerstein-Kozlowska et al. [18], it appears that a change in the oxide composition at 1.1 V occurs as the oxide goes from PtOH to PtO at this potential value. This could explain, in fact, the break in the

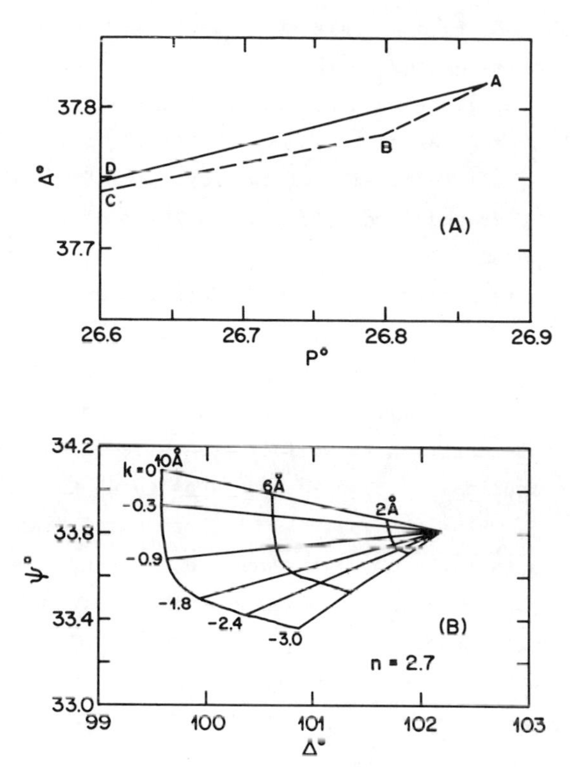

FIG. 14. (A) Experimental ellipsometer analyzer (A°) and polarizer (P°) readings upon platinum oxidation (ABC) and reduction (DA) [42]. (B) Theoretical ellipsometric curves for assumed refractive index n and variable absorption coefficient k; lines for definite thickness indicated [87], ψ is related to the amplitude ratio of elliptically polarized light

curve in Figure 14(A) but leaves the cathodic reduction curves unexplained. Horkans et al. [100] have observed a definite change in the optical constants (n and k) at 1.1 V and relate these changes to the formation of PtO from PtOH or from a bridged platinum oxide (two platinum atoms per oxygen) to PtO.

As mentioned previously, at high anodic potentials and for long polarization periods, another type of oxide called type II or β oxide is formed [21-32,56-62]. This second type of oxide has also been studied by ellipsometry [102,103]. The oxide formation at high anodic potentials (up to 2 V) has also been studied by Parsons and

Visscher [101] who observed a limiting value in the optical parameters (Fig. 15) corresponding to the limiting coverage observed coulometrically by Biegler and Woods [23]. The limiting coverage by the α oxide has been calculated to be 8 Å [101] or 13 Å [102] and is postulated [102] to consist of two layers of PtO and not, as proposed by Biegler and Woods [23], a monolayer of PtO_2. As the polarization potential and time increase, the formation of the type II or β oxide is observed and can be detected optically [102]. In Figure 16(B), the optical parameters (Δψ and ΔΔ) for the formation and reduction of the two oxides are shown for an oxidation potential of 2.2 V and a polarization time of 180 min [102]. Figure 16(A) shows the reduction potential-time curve for a constant cathodic current applied to the electrode subsequent to its anodic polarization [102]. From the analysis of these data, Vinnikov et al. [102] conclude that the α oxide reaches a constant thickness of 13 Å and the β oxide continues to grow as a function of time, up to a calculated thickness of 60 Å. However, above 2.3 V the rate of increase of thickness is slower, as was also observed previously [32].

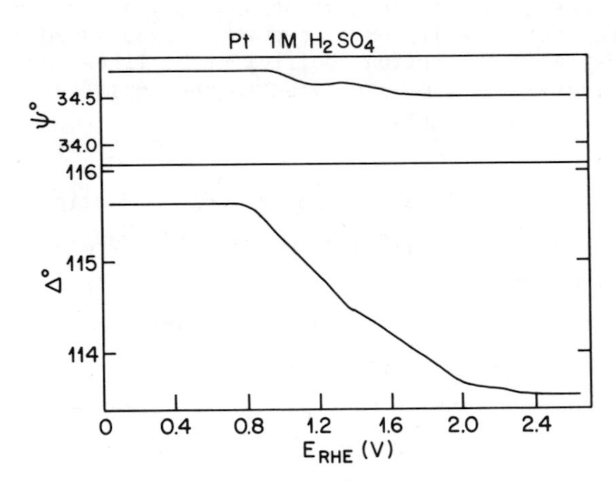

FIG. 15. Ellipsometric parameters indicating limiting oxide coverage, after Parsons and Visscher [35]. Platinum electrode in 1 M H_2SO_4.

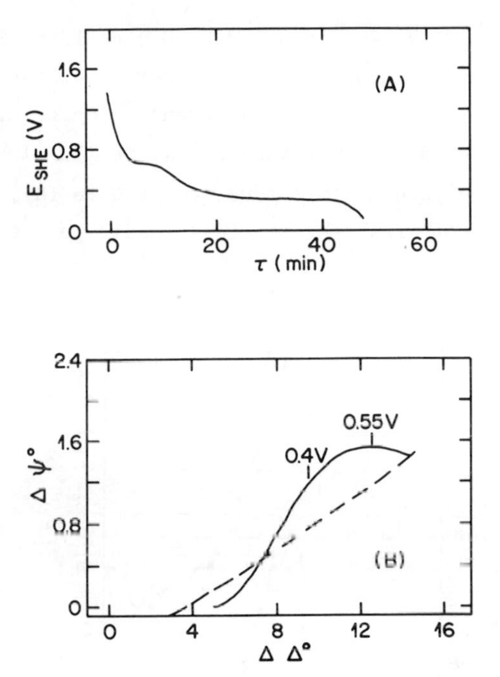

FIG. 16. Electrochemical and ellipsometric curves for type II oxide or β oxide [102]. (a) Galvanostatic reduction (6.2 μA cm^{-2}) of platinum electrode polarized at 2.2 V for 180 min. (B) Ellipsometric data for oxide formation and oxide reduction are shown by the broken and full lines respectively.

Vinnikov et al. [102] found it best to explain their results by the formation of a thin layer of α oxide, and underneath this layer they postulate the presence of the β oxide which can grow to relatively high thickness. They find a refractive index for the β oxide similar to the one observed by Visscher [96] for the $PtO.H_2O$ bulk oxide.

b. Other Optical Techniques. A technique related to ellipsometry is the electrochemical reflectance measurement. In this technique the potential of the electrode under study is modulated by a sinusoidal or a square wave perturbation. The relative change in reflectivity of polarized light is measured by a lock-in amplifier system synchronized to the modulating frequency. In one such study [104] on platinum, the reversibility of the oxide formation at potentials below 0.85 V (RHE in 0.5 M H_2SO_4) could be illustrated. As the potential of the anodic sweep is increased, the irreversible component of the system is clearly apparent. The results of Conway and Gottesfeld [104] are illustrated in Figure 17. In this figure the anodic scanning potential is reversed at 0.86 (A), 0.95 (B), and 1.12 V (C) and we note that for the less anodic potential scan the

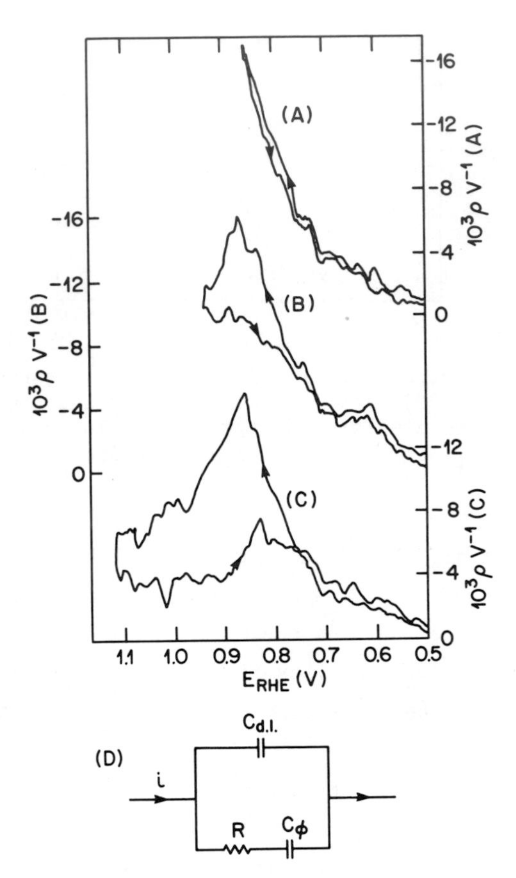

FIG. 17. Electromodulated reflectance data and equivalent circuit representation [104]. (A) to (C) ρ function defined as $(\Delta R/R)/\Delta V$ versus potential for different anodic limits: (A) 0.86 V, (B) 0.95 V, and (C) 1.12 V; v = 5 mV sec^{-1}, modulation frequency = 32.5 Hz, amplitude = 100 mV peak to peak. (D) Equivalent circuit; C_ϕ represents the pseudo-capacitance of Pt-OH adsorption, and $C_{d.l.}$ the double layer capacitance.

reversibility of the ρ function $(\Delta R/R)(1/\Delta V)$ is complete. However, as the potential is made more anodic, an irreversible component sets in. A similar effect is observed [104] when the scan is stopped at different potentials, i.e., stopping the scan below 0.9 V for 60 sec does not change the value of ρ for the cathodic sweep. For potential values higher than 0.9 V, the ρ function decreases during the arrest indicating an irreversible component of the system. Similar results have also been obtained by Barrett and Parsons [105]. On the basis of these results, Conway and Gottesfeld [104] further justify the proposed model [18,19] discussed earlier, in that in the potential range 0.72 ± 0.05 V, the reversible water discharge on a free platinum site is proposed as

$$Pt + H_2O \rightleftarrows PtOH + H^+ + e \qquad (1)$$

for which they suggest the equivalent circuit illustrated in Figure 17(D) to account for the measured capacitance curves [104].

As the potential is increased to values above 0.86-0.90 V, another electrode process occurs to account for the decrease in ρ and the measured capacitance. This change in mechanism occurs at low coverages of the platinum by PtOH, as at 0.9 V the coverage is of the order of 10-15% in PtOH. The formation of a less reversible PtOH species could be responsible for the observed behavior. The rearrangement of the PtOH as discussed previously could account for the decrease in ρ and C. The following scheme is proposed [104],

$$Pt \underset{}{\overset{I}{\rightleftharpoons}} PtOH \xrightarrow{II} HOPt \qquad (30)$$

in which stage II becomes important at potentials above 0.9 V. The same conclusion was reached from potential sweep experiments [104] in which the potential was held for different time periods (30 or 120 sec) at 1.06 V. After the holding period, the sweep was continued to 1.4 V and it was observed that the cathodic sweep after the holding period was identical to that pertaining to the repetitive sweep. This was explained [104] by the suggestion that the holding period has the same effect as an increase in potential, i.e., a rearrangement of the deposited PtOH layer. These experiments [104] and other work [18] elucidate the beginning of the oxide formation on platinum. The potentials studied do not explore, however, the formation of higher phases in the oxidation process; the limiting oxide coverage and the nature of the type II oxide or β oxide was not discussed in these investigations [18,19,104].

Barrett and Parsons [105] have drawn a similar conclusion from their study on the modulation of the ellipsometric parameter Δ. They observed a peak at 0.9 V, similar to the one in the reflectance experiments, for the curve $\partial R/\partial E$ versus potential. This value falls to zero at 1.25 V (versus SHE). This would seem to indicate an irreversible process but as the reflectance modulation ρ does not drop to zero [104,105] this would also seem to indicate a certain degree of reversibility in the oxide formation. The discrepancy between these results was not explained [105].

Differential reflectance of platinum electrodes as a function of potential [92,106] and as a function of the wavelength [106-108] of the incident light has also been measured. These results are depicted in Figures 18 and 19. The term $\Delta R/R$ is defined as $[R(d) - R(O)]/R(O)$ where $R(d)$ is the reflectance in the presence of a film and $R(O)$ in absence of such film. In the case of platinum, $R(O)$ is taken at 0.4 V [106]. In Figure 18, the variation of the differential reflectance with potential describes the familiar coulometric pattern. An important point to mention is the complete recovery of the original experimental points as the potential cycle is completed. This recovery indicates the absence of a residual film at the termination of the cathodic cycle. From their results, McIntyre and Kolb [106] postulate the presence of the oxide as a PtO species with the refractive index equal to 2.8 at 5461 Å with the absorption coefficient equal to 0.7. These values compare well with those obtained from ellipsometric measurements [87,96]. The linear variation of the differential reflectance $\Delta R/R$ with the wavelength (Fig. 19) from 2 to 4 eV can be deduced analytically [109] and is given [106,109] by

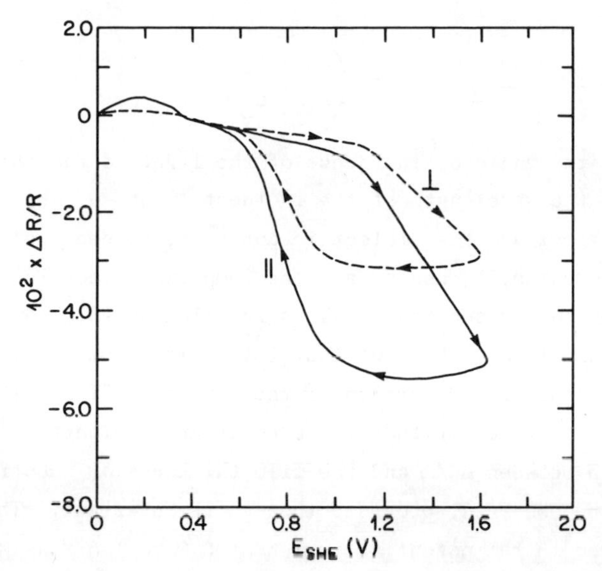

FIG. 18. Differential reflectance as a function of platinum electrode potential in 1.0 M H_2SO_4 at a wavelength of 3000 Å; v = 30 mV sec^{-1}; ⊥, perpendicular polarized light; ||, parallel polarized light; under Ar-saturated atmosphere; light 45° incidence [106].

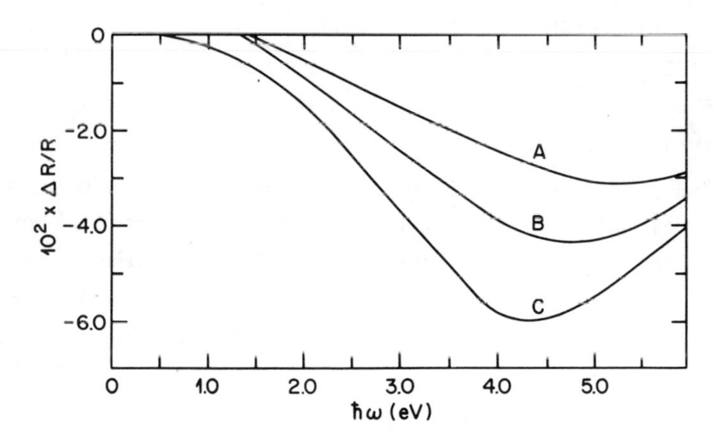

FIG. 19. Normal-incidence differential reflection spectra of platinum in 1.0 M $HClO_4$ for different electrode potentials as a function of incident light photon energy. A, 1.2 V; B, 1.4 V; and C, 1.8 V [106].

$$\left(\frac{\Delta R}{R}\right)_s = \frac{8\pi d n_1 \cos \phi_1}{\lambda} \operatorname{Im}\left(\frac{\hat{\varepsilon}_2 - \hat{\varepsilon}_3}{\varepsilon_1 - \varepsilon_2}\right) \tag{31}$$

where ϕ_1 is the angle of incidence of the light, d the thickness of the film, λ the wavelength of the incident light, n_1 and ε_1 are the refractive index and the dielectric constant, respectively, of the electrolyte medium, $\hat{\varepsilon}_2$ and $\hat{\varepsilon}_3$ are the complex dielectric constants of the film and the substrate. This relation applied for $d \ll \lambda$, i.e., for thin films. For constant film thickness, Eq. (31) takes into account the linear portion of the curves in Figure 18. From the equation it is clear that for a constant wavelength, a linear relationship between $\Delta R/R$ and the film thickness or substrate coverage will be observed, provided that ε_2 is invariant. This is also observed in the potential region of 0.9 to 1.6 V as seen in Figure 18. The minimum in the $\Delta R/R$ versus $\hbar\omega$ curve (Fig. 19) is interpreted as the formation of a second type of oxide, but the authors did not elaborate on the possible nature of this second type of oxide. In the above study [106], the light was modulated mechanically by a chopper; the potential of polarization was not modulated as in the other work cited previously [104,105,107,108].

The overall reflectance-potential curve obtained by Bewick and Tuxford [108] differs from the one reported by Conway and Gottesfeld [104] and Barrett and Parsons [105] in that a well-defined peak is not observed in the region of 0.8 to 0.9 V for the anodic sweep: the different type of modulation (square wave) used by the former authors [108] could possibly account for the difference. However, they did observe a relation between the incident wavelength and reflectivity similar to that observed by McIntyre and Kolb [106].

Another type of optical experiment was performed by Vinnikov et al. [103] to study the different (α and β) oxides. They measured the photocurrent obtained upon illumination of different oxides. For a monolayer of chemisorbed oxide (monolayer of PtO) formed by holding the platinum electrode at 1.4 V (versus RHE) in 0.5 M H_2SO_4 the photocurrent recorded is of a square-wave type:

rapid rise and then a constant value (see Fig. 20, I). As the coverage by the α oxide is attained, a high surge of anodic photocurrent is observed as the light strikes the electrode surface (Fig. 20, II). However, as the β oxide is formed, the anodic photocurrent decreases and as the thickness of the β oxide is increased, this photocurrent becomes cathodic (Fig. 20, III-V). The α oxide is monitored during illumination by ellipsometry and it is observed that the α oxide is reduced by about 4.8% upon illumination. The mechanism proposed for the decrease in α oxide is (in the absence of β oxide)

$$\underset{\alpha\text{ oxide}}{PtO[O]} + H_2O \xrightarrow{\hbar\nu} PtO + O_2 + 2H^+ + 2e \qquad (32)$$

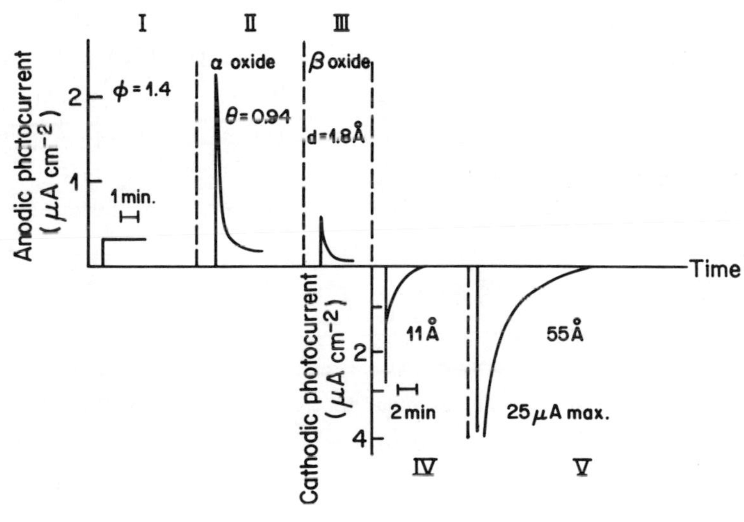

FIG. 20. Variation of photocurrent with time for a platinum electrode for different oxidation states [103]. I, electrode polarized at 1.4 V with a monolayer of chemisorbed oxygen present; II, electrode at 2.2 V until α-oxide coverage reaches 0.94; III to V, electrode polarized at 2.2 V until α-oxide coverage equals 1 and β-oxide thickness is as indicated [103].

In the presence of β oxide the proposed reactions upon illumination are

$$PtO_n + PtO[O] \xrightarrow{\hbar\nu} Pt^+O_n^- . PtO[O] + 2H^+ + 2e \rightarrow PtO_n + PtO = H_2O \quad (33)$$

β oxide charge transfer complex

The β oxide acts as a sensitizing substrate in the reduction of the α oxide. These results, as well as the work related to the β oxide (or type II oxide), are rather intriguing and deserve further careful study by all the available electrochemical and optical methods.

c. Spectroscopic Methods. In recent years, the chemical nature of the electrochemically-formed different oxides have also been examined by modern X-ray techniques (LEED, ESCA, and Auger). These techniques are able to give information as to the nature of the chemical species at the surface of a bulk material.

Kim et al. [110] were the first to report the ESCA spectra of anodically-treated platinum electrodes. More recently a complete analysis of the anodically-produced platinum oxides based on this method was given by Allen et al. [111]. The spectra obtained are shown in Figure 21 (a-e). As the anodic polarization becomes more drastic, new surface species are detected. The peaks associated with the strongly oxidized surface (Fig. 21 c-e) were attributed to the species PtO_2 [110-114].

These results of Allen et al. [111] differ slightly from those of Kim et al. [110] for the so-called oxide I. Kim et al. attribute the observed peaks to oxygen adsorbed from the atmosphere. They also observed additional peaks which they concluded to be associated with PtO; these peaks were not observed by Allen et al. [111]. These discrepancies perhaps arise from the different methods of producing oxides in the two studies. Kim et al. [110] polarized their electrode at 2.2 V for 3 min only, after a treatment in aqua regia and a reduction by $FeSO_4$. For Allen et al. [111], the high current density at

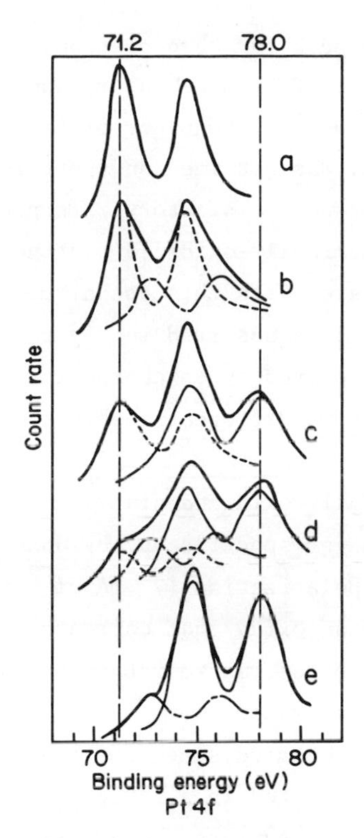

FIG. 21. ESCA spectra of platinum electrode under different anodization conditions [111]; a, electrochemically cleaned in 0.5 M H_2SO_4; b, at 2.4 V for 15 min; c, 4 V at a current density of 1 A cm^{-2} for 1 hr; d, same as c for 12 hr; e, same as c for 48 hr. Heavy solid lines represent the experimental curves, dotted lines give the deconvulated spectra (--- platinum metal; —— and — - — lines represent the oxidized species).

high anodic potentials produced a real phase oxide: their conditions are similar to those used by other workers to obtain the oxide of type II or β oxide [29-32,56-62]. Allen et al. attributed their oxide I to the species $PtO.2H_2O$ or $Pt(OH)_2$ but the evidence for these species are of electrochemical nature only.

Dickinson et al. [112] contested the interpretation of Allen et al. [111] in particular, the observed peaks attributed to the platinum metal. Dickinson et al., by a proper deconvolution of the experimental spectra, observed satellite peaks on the high energy side of each major 4 f peak for bare platinum (at 71.2 and 74.5 eV). In fact the single peaks of Figure 21a are doublets. This interpretation is not borne out in the paper of Bancroft et al. [113] where the ESCA spectrum was obtained on sputtered platinum films. At the moment, there is no satisfactory interpretation of the ESCA spectrum for electrochemically-oxidized platinum electrodes. One of the problems associated with this technique arises from the absence of definite peaks: as observed in Figure 21, the experimental peaks have to be deconvoluted by an appropriate technique and this procedure may give rise to different observations and interpretations.

The electrochemically-oxidized platinum electrodes have also been investigated by Auger spectroscopy by Johnson and Heldt [115]. For a relatively mild polarization (77 mA cm^{-2} for 16 hr) they detected the presence of an oxide that corresponds approximately to the PtO species. This oxide was resistant to an electron beam bombardment in the spectrometer.

Shibata [116] has reported some electron diffraction studies on platinum electrodes treated in a variety of ways. He investigated mainly the change in surface characteristics brought about by anodic-cathodic cycling experiments. When an electrode was anodized for 3 hr at 200 mA cm^{-2}, a yellow colored film formed on the electrode; its composition was found to be PtO_2, as identified by the LEED data of a chemically-prepared sample. When this oxide was reduced cathodically, the surface acquired the character of platinum black or the electrochemically-deposited platinum. To observe these changes in surface structure the polarization had to be performed at potentials above 1.15 V. It is only above this potential that the surface oxidation exceeds one oxygen atom per platinum atom in the form of PtOH. It is at this potential that the place-exchange mechanism starts to operate and the oxygen can penetrate into the metal [18,19]. For electrodes treated under milder conditions Schubert et al. [117] failed to observe any significant change in their LEED pictures. The anodic polarization ranged from 0.5 to 2.2 V for up to 80 min but no clear evidence for any surface oxide could be deduced from these data.

C. Conclusions

Despite numerous recent studies based on a variety of experimental procedures, a clear and definitive picture has not yet been obtained of the anodic oxides formed on platinum under various conditions, although a great deal of progress has clearly been made. The controversy regarding the nature of the anodic oxides on platinum and the mechanisms for their formation is still as intense as it was a few years ago.

A general summary that tries to reconcile the different results and attempts to subdivide the problem into its components is presented below.

Starting from a platinum surface free of oxides, in acidic solutions, as the anodic potential is increased water will be discharged to give a PtOH species; this discharge is reversible up to approximately 0.9 V (versus SHE). Above this potential, rearrangement of the deposited PtOH takes place, which introduces irreversibility in the system, and the familiar hysteresis between oxide formation and reduction occurs. The PtOH formation and its rearrangement to give OHPt occurs via a place-exchange mechanism according to the following reactions:

$$Pt + H_2O + PtOH + H^+ + e \rightarrow OHPt \quad (34)$$

As the potential increases further, complete coverage is attained at 1.1 V ($Q_O/Q_H = 1$). Above this potential the OHPt species can be oxidized further to give the PtO oxide. As the potential is increased to above 2 V, a limiting coverage in the oxide is obtained. The stoichiometry of the oxide now corresponds to a structure where $Q_O/2^SQ_H = 2$. This is compatible with a monolayer of PtO_2 or a surface in which the PtO oxide is two layers thick. The latter picture falls more in line with the optical data obtained so far.

For more vigorous anodizations, i.e., high potentials for

extended periods of time (hours and even days), a new type of oxide is formed; the platinum becomes yellow in appearance and a new arrest develops in the galvanostatic reduction at an overvoltage higher than that associated with the arrest observed for the reduction of the oxide formed at lower anodic potentials. This second type of oxide increases in thickness up to 60 Å and does not show any limiting thickness behavior. The maximum rate of formation is at about 2.1-2.3 V. The nature of this oxide has been clearly defined as the PtO_2 species by spectroscopic methods.

The mechanism of the PtO formation is not yet clearly defined. Even the experimental data are not completely above some doubts. The logarithmic law for the oxide growth with time is observed generally but the time scales used by different authors are not the same. This logarithmic law was observed for a complete range of time from milliseconds to 1,000 sec by Vetter and Schultze [13] but only for a more restricted time range by Gilroy and Conway [22] (from around 20 to 1,000 sec).

The variation of Tafel slope for oxide growth with the oxide coverage of the electrode has led to the suggestion of the high-field ion conduction mechanism in oxide growth [7,13-15,42,118]. However, this mechanism cannot explain the limiting coverage observed by many workers. Also, the variation of the Tafel slope (b) with coverage can be accounted for by a completely different mechanism, as shown by Nadebaum and Fahidy [119]. The different values of b for different coverage values, determined at constant θ, can be obtained for the mechanism involving the following adsorption-desorption reaction [119]:

$$Pt + H_2O \rightarrow PtOH + H^+ + e \qquad (35)$$

for which the rate equation can be written as

$$\frac{d\theta}{dt} = k\, C_{H_2O}\, (1 - \theta)\, \exp\, [(1 - \beta)(FV/RT - m\theta)] - k^{-1}\, C_{H^+}\, \theta\, \exp\, [-\beta(FV/RT + m\theta)] \qquad (36)$$

For constant θ, this expression reduces to

$$V = b\ X\ (\text{constant}) \tag{37}$$

The high-field ionic conduction growth could probably be applied to the oxide formed at high potentials (>2 V), namely, the PtO_2 oxide, whose formation cannot be explained by the above treatment. Nadebaum and Fahidy [119] also examined other mechanisms and they find oxide growth by incorporation of oxygen atoms into the metal the most plausible mechanism to account for their results in acetate solution [119]. In the above mechanism, reaction (35) is assumed to be at equilibrium with subsequent steps containing the r.d.s. The r.d.s. is postulated [119] to be without a charge transfer and so is unaffected by the electrode potential in all cases of oxide growth; this assumption, however, is contestable in several cases of oxide growth. The final and definitive conclusions on the subject cannot yet be deduced in view of the variety of experimental facts and mechanistic proposals which in many cases are not reconcilable to each other.

III. THE ANODIC OXIDE FILM ON GOLD ELECTRODES

A. Introduction

As stated previously, other noble metals have received much less attention than platinum. However, gold has been the subject of several electrochemical studies in the past, and recent optical techniques have also been applied to the oxide formation on this metal.

The literature published prior to 1965 has been covered in great detail by Hoare [1] and only the general features of this work will therefore be discussed here. We will concentrate our attention mainly on the more recent investigations on the subject.

The anodic polarization of gold electrodes in acidic solutions leads to the formation of an oxide on the electrode; this oxide has now been well characterized as the Au_2O_3 or its hydrated species. The Au-Au_2O_3 system is well defined and the reversible potential set up by the following equilibrium has been measured [120].

$$Au_2O_3 + 6H^+ + 6e \rightleftarrows 2Au + 3H_2O \qquad (38)$$

$$E_0 = 1.36 \text{ V}$$

The onset of this oxide formation occurs at 1.35 V, as determined from the potential sweep experiments [121]. The reduction and oxidation peaks show, as in the case of platinum, a high degree of hysteresis and the current maximum of the reduction peak is located around 1.1 V [121]. The anodic and cathodic charge ratio for these experiments and for galvanostatic charging curves is equal to 1. The Au_2O_3 oxide is a poorly adherent species, has a flaky nature, and is highly colored (reddish brown to black) [1].

In alkaline solutions, no such formation of a phase oxide [Au_2O_3) is observed; the electrode keeps its golden appearance and, at best, a monolayer of such oxide is formed [1]. However, the anodic dissolution of gold is more important in alkaline solutions.

Fine structures have been observed by some authors in the galvanostatic charging curves of gold, especially at very low current densities [122]. These structures or plateaus were attributed to other forms of oxides, namely Au_2O and AuO. However, the possibility of reactions of impurities cannot be ruled out since the time scale for observation was rather long (up to 3 hr).

In the following sections, the recent electrochemical and optical studies on the anodic oxide formation on gold will be described.

B. Nature of the Anodic Oxides on Gold and their Mechanisms of Formation

1. Electrochemical Evidence

Brummer and Makrides [123] and Brummer [124] have reported on the kinetics of anodic oxide formation on gold electrodes. They oxidized the electrode potentiostatically and by the galvanostatic reduction of the so-formed oxide, they measured the quantity of oxide formed. The reduction profile consists of a main arrest occurring at a constant potential. They observed a linear increase in the charge, at constant time of formation, with the formation potential. The results agree with the previous work of Laitinen and Chao [125] at low potentials. They [125] did observe a variation in the reduction charge with the reduction current density, in solutions of pH values less than 1. The observation could perhaps be explained by a concomitant chemical dissolution but the corresponding dissolution current (0.3 μA cm^{-2}) is much too small compared with the experimentally-observed current deduced from the oxide reduction charge versus time curves. They [123] interpreted their results by postulating the decomposition of an active intermediate species during the electrochemical oxidation.

For the oxide reduction, they [123] found a Tafel relationship where the slope was of the order of 39-42 mV; this slope was invariant for different potentials of oxide formation and was independent of the pH (from 0.06 to 1.60) of the working solution. The pH-dependence of the reduction current at constant potential was found to be -1.39 as defined by $[\partial(\log i)/\partial pH]_E$.

To account for these observations they proposed the following mechanism:

$$AuOOH + H^+ + e \underset{}{\overset{fast}{\rightleftharpoons}} AuO + H_2O \qquad (39)$$

$$AuO + H^+ + e \xrightarrow{slow} AuOH \qquad (40)$$

$$AuOH + H^+ + e \xrightarrow{fast} Au + H_2O \qquad (41)$$

Solving for [AuO] and substituting the value in the appropriate rate equation, they [123] obtained the following rate law:

$$i = k[AuOOH][H^+]^2 \exp - \frac{(1 + \alpha)}{RT} \qquad (42)$$

where [AuOOH] is the gold oxide coverage (the hydrated form of Au_2O_3) and k is a constant. Equation (42) gives rise to a Tafel slope $[\partial E/\partial(\log i)]$ of 39 mV for $\alpha = 0.5$ and a pH-dependence $[\partial(\log i)/\partial pH]$ of -2. These values show a fairly good agreement with the experimentally-observed data. A better agreement for the pH-dependence could be obtained by invoking cation adsorption on the electrode [123]. Brummer and Makrides reported an experimentally-determined pH-dependence of -2.0 when an excess of an inert salt [$Mg(ClO_4)_2$] was added to the solution for pH values of 0.2 or less, and, at high oxide formation potentials.

In the second study, Brummer [124] looked into the ageing effects of the oxide. He reported a logarithmic oxide growth law and the rate of growth [$\partial Q/\partial(\log \tau)$] increased as the potential of oxide formation increased. As the time of oxide formation increases, the oxide becomes harder to reduce (the potential of reduction is shifted to more cathodic values). This indicates an important time effect which, according to Brummer [124], excludes the use of the cyclicvoltammetric method of Will and Knorr [121]. It may be pointed out, however, that the cyclicvoltammetric method can give information as to the first steps of the oxide formation. As with platinum, the gold oxide undergoes a rearrangement into more stable forms at higher anodic potentials so that important time effects are observed.

In recent publications, Schultze et al. [126] and Dickertmann et al. [127] studied the kinetics of the oxide formation and dissolution on gold (polycrystalline and single crystals) electrodes in an acidic medium. The techniques used were mainly the anodic and cathodic galvanostatic charging curves [126] and potentiodynamic methods [127]. A typical charging and reduction curve is shown in Figure 22. The reduction current is switched on at different potentials on the anodic charging curve. From the cathodic arrest the oxide coverage of the electrode can be estimated. For gold it is impossible to use the normal procedure for the determination of oxide coverage on platinum because hydrogen is not adsorbed on the gold electrode (${}^{S}Q_H$ cannot thus be measured). To calculate the coverage Schultze et al. [126] estimated the charge needed to attain a monolayer of oxygen on the gold surface; such a charge for the (100) crystallographic plane is 380 $\mu C\ cm^{-2}$. However, a roughness factor has also to be defined to take into account the active sites of the electrode and so the coverage is then defined as

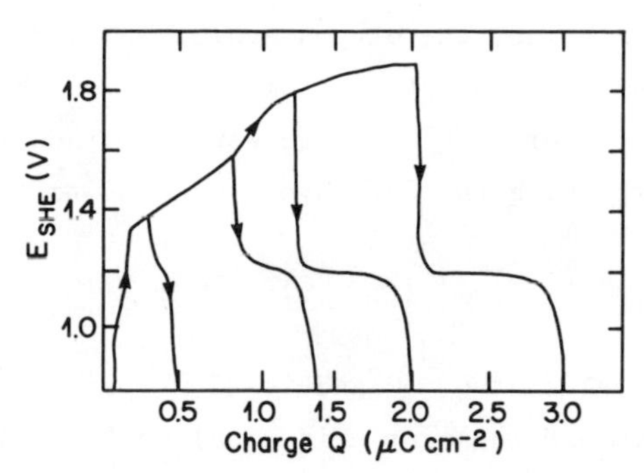

FIG. 22. Anodic galvanostatic charging curve for a gold electrode (40 μA cm^{-2}) in 0.5 M H_2SO_4 at 25°C; when different potentials (1.4, 1.6, 1.8, and 1.9 V) are reached, the galvanostatic oxide reduction (same current density) is carried out [126].

$$\theta = \frac{Q_{measured}}{r \times 0.380} \tag{43}$$

where r is the roughness factor and $Q_{measured}$ is the experimentally measured charge associated with cathodic arrest (see Fig. 22).

These authors [126] determined a Tafel-type relationship at different values of constant coverage. The results are shown in Fig. 23. The oxide formation is the only reaction for coverages up

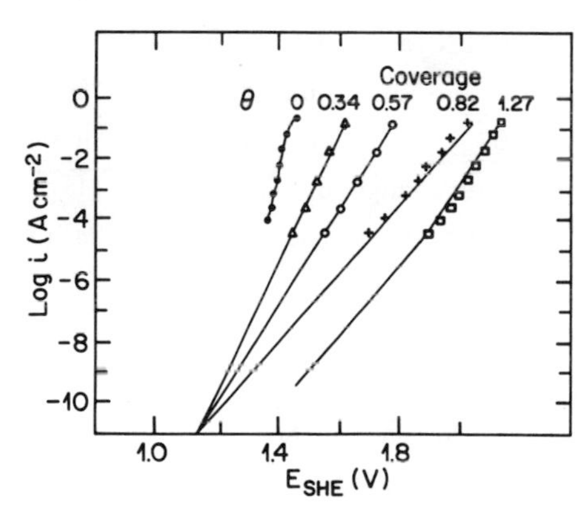

FIG. 23. Tafel plots for the oxide formation on a gold electrode at different constant oxide coverages from 0 to 1.27 in 0.5 M H_2SO_4 at 25°C [126].

to 0.82; above this value, the oxygen evolution reaction will complicate the analysis of the data. From the curves shown in Figure 23, the Tafel slopes for each coverage were obtained [126]. These slopes vary linearly with the coverage according to the following law:

$$b = b^{o}(1 + a\theta) \tag{44}$$

where b^0 is the b value at $\theta = 0$ and is 25 mV per decade. The experimentally-determined constant a is equal to 3.0. The log i versus potential relationship can then be written as

$$\log i = \log i_0 + \frac{\varepsilon - E}{b^{o}(1 + a\theta)} \tag{45}$$

where $\log i_0$ is a constant equal to 2×10^{-11} A cm^{-2} and E equals 1.15 V; these values were obtained by the extrapolation of the experimental lines as shown in Figure 23.

The variation of the Tafel slopes with coverage is similar to the behavior on platinum as was discussed previously. This is interpreted as a manifestation of the high-field ionic conduction mechanism. As the oxide layer grows, a part of the potential will be lost in the oxide layer and the field at the metal oxide-electrolyte interface will thus be lower so that a higher overpotential for the oxide formation will result. The equivalent-circuit representation used by Schultze and Vetter for gold [126] was the same as that for platinum. The same mechanism is also postulated in which the oxide ion (O^{2-}) is in equilibrium with water and the rate r.d.s. is in the place-exchange between a metallic ion at the metal-electrolyte interface and the oxide ion. As the layer grows, the slowest step could occur at the metal-metal oxide interface or at the metal oxide-electrolyte interface.

The reduction behavior of gold oxide has also been studied. For this process, a variation in the Tafel constant with the oxide coverage was observed but the magnitude of the variation was less pronounced; the variation was from 30 mV per decade to 43 mV per decade for coverages ranging from 0 to 1.27. These values compare favorably with those obtained by Brummer and Makrides [123]. From this small variation, Schultze and Vetter [126] concluded that the oxide reduction proceeds at the edges of islands of oxide. As the coverage decreases this reduction can eventually proceed uniformly.

The same conclusion, namely the high-field ionic conduction mechanism, was also deduced from the electrochemical study of oriented single crystals. Figure 24 illustrates the cyclicvoltammograms obtained by Dickertmann et al. [127] for different gold electrodes. We note the important differences between the single crystal with two different orientations, (100) and (111), and the polycrystalline electrodes. The anodic oxide formation is very sensitive to the crystallographic nature of the electrode. The oxide reduction peak is indifferent, however, to the type of substrate used. The kinetics of oxide formation depend very much on the nature of the surface, and Dickertmann et al. [127] concluded that

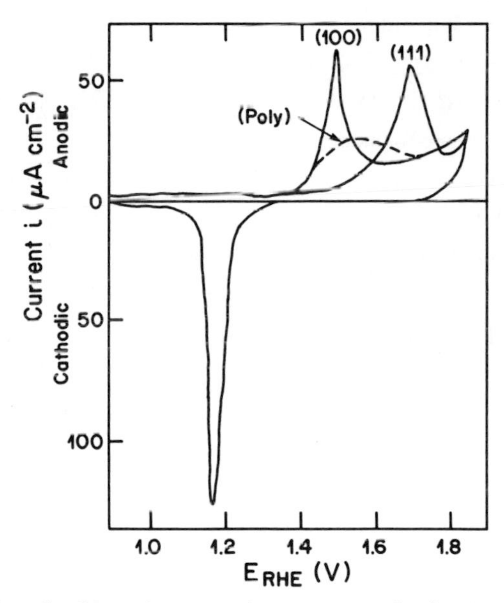

FIG. 24. Cyclicvoltammetric curves of single-crystal [(100) and (111) planes exposed] and polycrystalline gold electrodes in 1 M $HClO_4$ at 25°C; v = 10 mV sec^{-1} [127].

the differences in the rates of oxide formation are caused either by the variations in the effective charge of the reacting species or by the differences in the potential distribution between the planes. These differences are important for the initial oxide formation (at thicknesses less than 2-3 Å). As the thickness increases (>3 Å) the variations in mechanisms between the different types of surfaces vanish and the high-field-assisted ionic conduction mechanism sets in.

Similar cyclicvoltammograms were also obtained by Hamelin and Sotto [128] and Sotto [129] for single-crystal gold electrodes in neutral solutions.

The oxide formation and reduction has recently been examined by cyclicvoltammetric methods at different temperatures [130-132]; the typical voltammetric curves are shown in Figure 25. We note the onset of the anodic oxide formation at approximately 1.35 V (versus SHE). The oxide formation curve is resolved into three distinct and well-defined peaks; the relative heights of these peaks are very much dependent on the temperature and also on the electrolyte concentration. At 25°C, for an acid concentration of 1 M, Capon and Parsons [130] did not find such a structure. The exact assignment of the peaks was not made by Ferro et al. [131] but an explanation similar to that for platinum [18,19] could possibly account for the peaks; one may invoke the possibility that different planes of the gold surface change with the temperature. The cathodic reduction of the oxide shows a very narrow single peak at 70°C. As the temperature is lowered to -11°C, two peaks are then observed [131]; the voltammetric curve for this temperature is shown in Figure 25(B). The parallelism between the platinum and the gold oxidation behavior cannot be extended too far since the voltammetric peaks do not show the same behavior as the sweep rate is varied. For platinum, the three anodic peaks are independent of the sweep rate [18]; for gold, however, these peaks are shifted anodically with increasing v (sweep rate), as is also observed for the single crystal gold electrodes [127]. The slope of ΔE versus

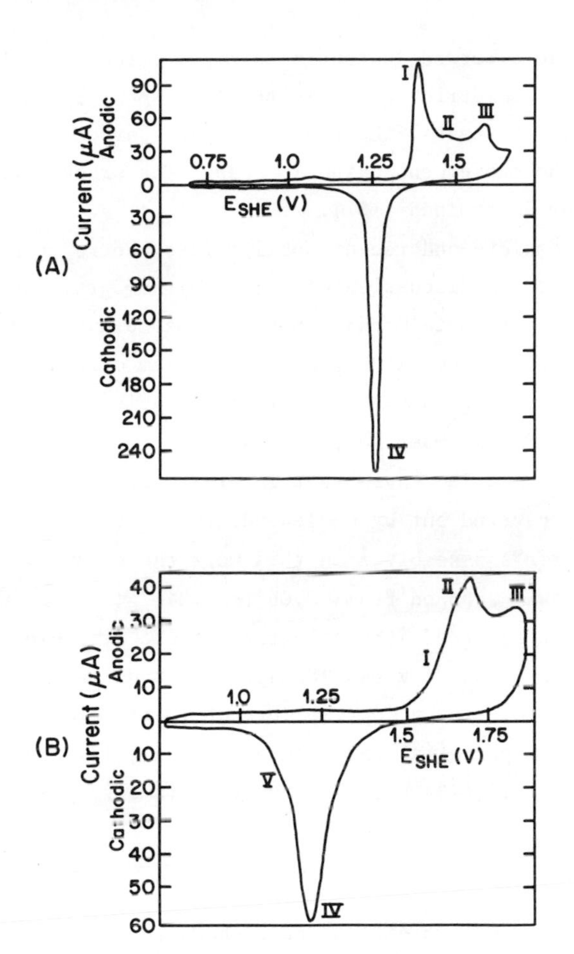

FIG. 25. (A) Cyclicvoltammetric curve of a polycrystalline gold electrode at 75°C in 5 M H_2SO_4; v = 40 mV sec^{-1} [131]. (B) Cyclicvoltammetric curve of a polycrystalline gold electrode at -11°C in 5 M H_2SO_4; v = 40 mV sec^{-1} [131].

log v yields values of the order of RT/2F (about 30 mV at 25°C) [131].

An additional fact observed by Ferro et al. [131] is the chemical dissolution or desorption of the oxide at open circuit. This dissolution was suggested by the decrease in the magnitude of the cathodic current peak for the oxide reduction, observed following an arrest of the electrode potential at open circuit at the anodic limit of the voltammetric scan. As the time spent by the electrode at open circuit increases, the reduction peak current correspondingly decreases in the subsequent cathodic scan. The exact nature of this process was not determined [131].

In more complete and recent publications, Ferro et al. [133, 134] gave a thorough discussion of cyclicvoltammograms obtained under various experimental conditions. The conclusions reached are similar to those for the oxide growth mechanism of platinum [18, 19]. In the case of gold the similar rearrangement of the anodically-formed oxide is observed [133,134].

However, a detailed investigation of dissolution of gold and gold oxide was carried out by Cadle and Bruckenstein [135] using a disk-ring electrode assembly. In this work the potential of the disk electrode was scanned from -0.06 to 1.84 V (versus SHE) and the ring current recorded; the potential of the ring electrode was held constant at 0.244 V (versus SHE). The resulting curve is shown in Figure 26. The reduction current at the ring was examined [135] as the disk potential was scanned cathodically from an oxide formation potential (1.84 V). The reaction involved is the gold deposition:

$$3e + Au(III) \rightarrow Au \qquad (46)$$

Cadle and Bruckenstein [135] have shown that the oxidation state of the soluble gold species was really Au(III). However, if the disk electrode is held at a constant anodic potential for 5 min, other interesting features begin to show up. As the potential of anodization increases, the ring current also increases. The symmetrical peak for the cathodic gold deposition current at +1.3 V becomes asymmetric as the anodic oxidation potential of the disk increases and finally a new peak at more anodic potentials is apparent as the potential reaches 1.7-1.8 V. This second peak is attributed to the species Au(I). The following reaction path was proposed [135] to account for the various experimental facts:

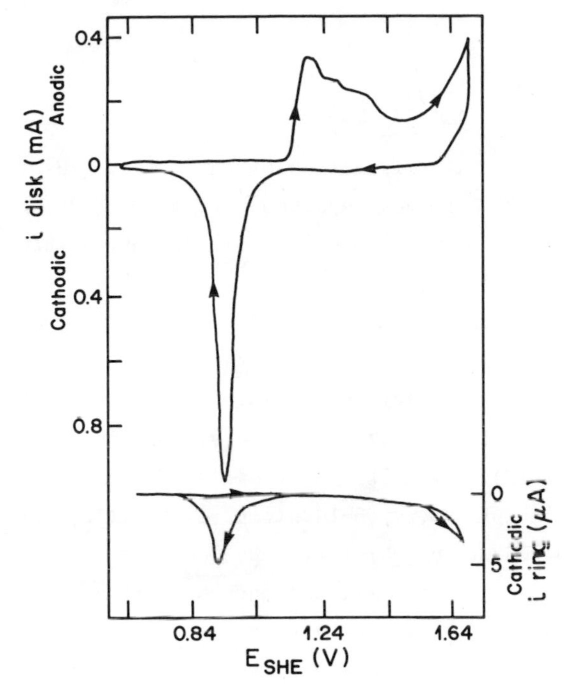

FIG. 26. Rotating disk voltammograms of a gold electrode in 0.2 M H_2SO_4, ω = 2,500 rpm; the gold disk was scanned at 100 mV sec^{-1} and the gold ring was held at a constant potential of 0.24 V [132].

$$AuO_x \xrightarrow{k(E)} Au(III) \xrightarrow{k(0)} AuO$$
$$\downarrow$$
$$\text{Solution}$$

where the production of Au(III) species is potential-dependent with a rate constant k(E) and its reduction to metallic gold has the rate constant, k(0); Au(III) can also be carried away in the solution by convection. The above mechanism gives a flux balance according to the following equation:

$$k(E) = k(0)C_s + L\omega^{1/2}C_s \qquad (47)$$

where C_s is the surface concentration of Au(III), ω is the disk rotation speed, and L is a group of constants from the Levich equation. Solving for C_s and substituting in the Levich equation yields [135]:

$$\frac{i_R}{N} = L\omega^{1/2}C_s = \frac{L\omega^{1/2}k(E)}{L\omega^{1/2} + k(0)} \tag{48}$$

The experimental confirmation of the $1/i_R$ versus $1/\omega^{1/2}$ relation as a straight line was reported. From Eq. (48) and the experimental data, the quantity of Au(III) diffusing into the solution can be determined [135].

The dissolution of gold was also examined by Rand and Woods [24] who found that the difference between the anodic and cathodic charges (Q_a and Q_c) in cyclicvoltammograms can be accounted for by the anodic dissolution of the noble metal (platinum, palladium, rhodium, and gold).

In a series of papers, Goldshtein et al. [136,137] investigated the oxidation-reduction behavior of gold oxides for very short times of oxide formation (in the milliseconds time range). For such short times of oxidation followed by a very fast reduction at a high sweep rate (i.e., 100 V sec^{-1}), they observed two to three peaks in the cathodic reduction of the oxide. Three peaks are observed in perchloric acid [137] but only two in sulfuric acid [136]. This structure in the cathodic reduction peaks is observed only for fast transients and also for relatively short periods of oxide formation. As this time is increased into the seconds range, the familiar singular reduction peak is generated. They [136,137] interpreted their results by invoking the presence of different types of sites on the gold electrode; as the oxide formation proceeds, the oxides with lower energy are converted into more stable forms of oxide. The evolution of these peaks is depicted in Figure 27 for an oxide formation potential of 1.45 V (versus RHE) and for different polarization times (from 1 to 100 msec). We may note the rapid decrease in the size of peak I (most weakly-bound oxygen) and the gradual increase in the heights of peaks II and III. The detailed analysis of the relative charges associated with these peaks shows [137] that the form I is converted into the form II whereas the form III reaches

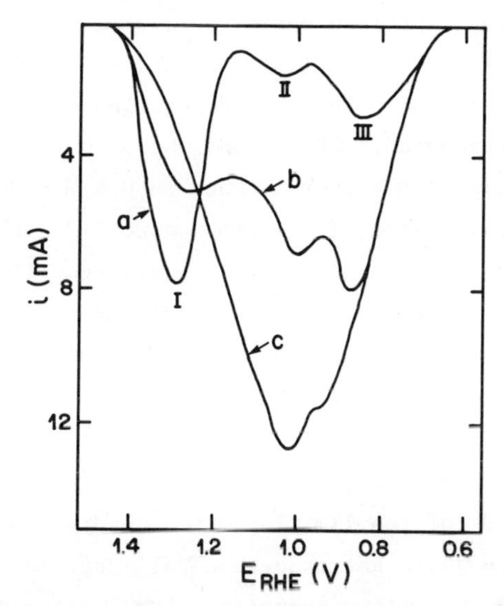

FIG. 27. Cyclicvoltammetric curve for the oxide reduction on a gold electrode in 2 M $HClO_4$; anodic polarization of 1.45 V for 1, 2, and 100 msec (a,b,c, respectively); v = 100 V sec^{-1} [137].

a constant limiting coverage (θ = 0.1 of a monolayer) and does not seem to be converted into other forms.

The conversion from the first type of oxide to the second is postulated [137] to follow the mechanism

$$Au + H_2O \rightarrow Au(OH) + H^+ + e \tag{49}$$

followed by

$$AuOH \rightarrow Au(O) + H^+ + e \tag{50}$$

The form III could then be another type of oxide of greater stability. However, no other evidence was presented to confirm the above mechanism so that the proposed reactions can be considered only as speculative.

No such resolution was observed in galvanostatic reductions [126] since the oxidation times used were sufficiently long to allow the rearrangement of the oxides to take place. The low-temperature results of Ferro et al. [131] where two cathodic reduction peaks were observed could be related to the same type of rearrangement of the oxide. Also Gruneberg [138] did observe the resolution of the cathodic peaks as a function of the anodic limit of the cyclicvoltammograms. For scans where the anodic limit is below approximately 1.75 V (versus RHE) in 0.5 M H_2SO_4, the cathodic reduction profile is resolved into two well-defined peaks. When this anodic limit is exceeded, only a single reduction peak is observed. This shows that not only are the time effects important but the formation potential of the oxide also plays a major role in the nature of the oxide formed. The importance of the potential of oxide formation was also pointed out by Schultze and Vetter [126] and Dicktermann et al. [127].

In their study of the limiting oxide coverage on noble metals, Rand and Woods [139] did not observe a well-defined limiting oxide coverage for gold but only a change of slope in the curves of reduction charge versus the oxidation potential. However, extrapolation of the onset voltage of phase-oxide formation leads to a roughness factor of 1.3 for an assumed monolayer of AuO species. As the anodic potential is increased to values above 2.0 V the gold electrode becomes covered by a deep orange layer (corresponding to the Au_2O_3 oxide); upon reduction the electrode retains a black appearance (probably fine particles of gold) and the surface roughness increases by a factor of 20 [139].

In one of the very few analyses of the anodic charging curves, Moslavac et al. [140] found a linear relation between the charging current density and the reciprocal of the transition time τ of the arrests in the galvanostatic charging curves. At low anodic currents, however, the linear relationship fails to be obeyed. To account for their experimental results, they [140] proposed a mechanism in which the simultaneous formation of an anodic layer and its dissolution are invoked. The dissolution of the layer is assumed to be dependent on the electrode oxide coverage. This treatment [140] leads to the following relationship:

$$\tau = -\frac{1}{a} \ln \left(1 - \frac{a}{b}\right) \tag{51}$$

where τ is the transition time and a and b are defined as

$$a = ki_k S^{-1} \tag{52}$$

$$b = ki_s S^{-1} \tag{53}$$

where k is the number of cm^2 covered by one coulomb, and i_k and i_s are the corrosion current and the anodic apparent current density, respectively, and S is a constant. For b >> a, the above can be expanded into a linear series and the current versus τ relationship becomes

$$i_s = \frac{1}{k\tau} + i_k \tag{54}$$

From the experimental curve, k and i_k were evaluated; with these values the i versus τ curve was calculated and found to agree with the experimental curve. One problem with the above treatment arises partly from the evaluation of the transition time τ. In the original paper, Moslavac et al. [140] did not state explicitly how they performed the anodic steps. No definite arrests were reported by Brummer and Makrides [123] in the anodic charging curves. An arrest is observed only during the oxide reduction [123]. The dependence of the arrest period on the current density for a fixed anodic charge yields a corrosion current of the order of 0.3 μA cm^{-2}, and not, as determined by Moslavac et al. [140], 290 μA cm^{-2}. The discrepancy probably arises from the method used [140] in the determination of this current density. Cadle and Bruckenstein [135] found an even lower rate of corrosion (0.035 μA cm^{-2}) from their ring-disk electrode study. One explanation put forward to explain the discrepancy in the corrosion rates is the possible presence of traces of chloride ions that will accelerate this corrosion [135].

Hoare [141,142] has identified two types of oxides on the gold surface. If a gold electrode that has never been oxidized electrochemically is examined in a sulfate solution purged by nitrogen gas, a double-layer capacitance of 22.4 μF cm^{-2} is found [141]. If the same electrode is now subjected to an oxygen atmosphere in the solution and then this oxygen is removed by nitrogen bubbling, the double-layer capacitance falls to 12.4 μF cm^{-2}, presumably owing to the presence of an oxide on the electrode. The original capacitance is recovered by cathodic removal of this oxide. According to Hoare [141] this indicates the presence of an oxide at a potential as low as 870 mV. This is contrary to the results of the cyclic-voltammetric experiments where the detection of oxide formation occurs at approximately 1.3 V. The partial oxide coverage responsible for the fall in capacitance is described [142] as a mixed electrode process involving the Au/AuO couple. Oxygen gas reduction and oxygen evolution were also studied on this type of electrode [142]; the latter reaction is impeded by the conversion of this type of oxide into the Au/Au_2O_3 couple [142].

If the gold electrode is oxidized at 2 V for several minutes, the double-layer capacitance subsequent to the electrochemical reduction of the oxide is very high (131 μF cm^{-2}). This sudden increase in capacitance is indicative of a surface roughening. Such surface roughening was also observed by Rand and Woods [139] for anodic polarization above 2 V. This increase in surface area presumably arises from the production of gold black (similar to platinum black) by the cathodic reduction of Au_2O_3. The conclusion that oxygen is adsorbed at about 900 mV [141] is in contradiction with several investigations [121-126,143]. It has been proposed [143] that no oxide is present below 1.20 V (versus RHE). The above discrepancy could be accounted for by the electrode pretreatment performed by Hoare [141] where the gold electrode was treated in a hydrogen flame and quenched in concentrated nitric acid. This last procedure could introduce oxygen into the bulk of the metal.

2. *Oxide Formation Studies by Nonelectrochemical Techniques*

As for the platinum electrode, the gold surface has also been studied by modern optical methods, particularly ellipsometry [94, 144-149] and reflectance methods [99,100,106,107,150,151].

Vinnikov et al. [146,147] have reported an extensive ellipsometric examination of gold oxide formation; their results are shown in Figure 28. They reproduced well the data of Sirohi and Genshaw [145]. The Δ-potential curve can be subdivided, in an acidic medium, into three regions. From 0.0 to 0.5 V, the ellipsometric data remain unaltered showing no oxidation of the metallic surface. From 0.5 to about 0.9 the Δ-value decreases gradually and, finally, above 1.2 V an abrupt decrease in Δ occurs. The Δ-potential behavior follows very closely the Q (oxide charge)-potential curve. The rapid decrease in Δ above about 1.2 V clearly corresponds to the oxide film formation. The familiar hysteresis between the oxide formation and reduction is also observed ellipsometrically [145]. The change in Δ in the potential range below 1.2 V was

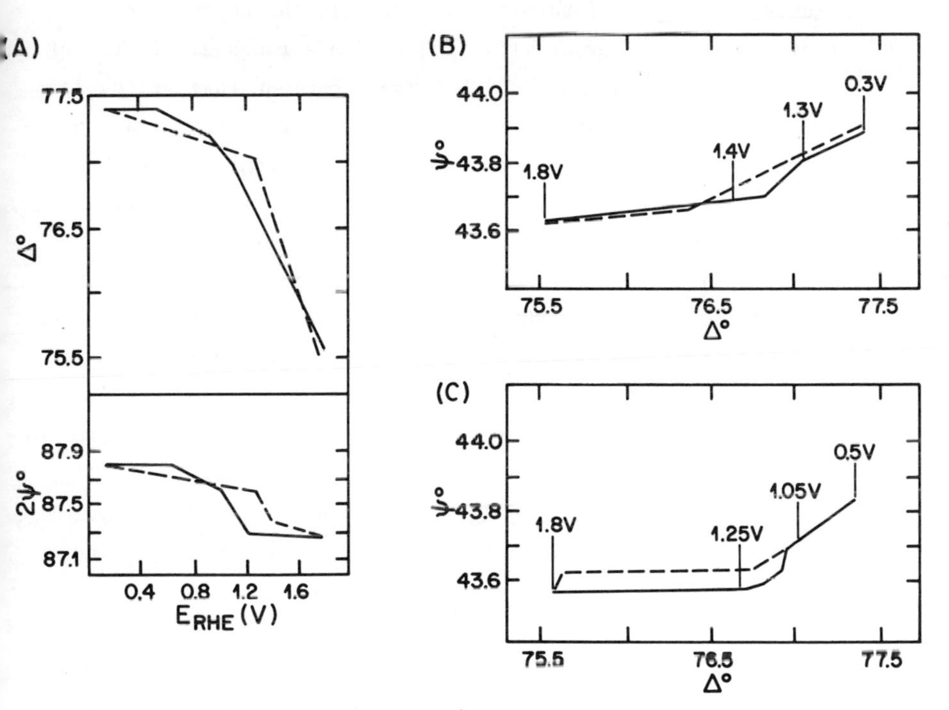

FIG. 28. Ellipsometric data for gold electrodes. (A) Variation of the ellipsometric parameter as a function of polarization potential for 0.5 M H_2SO_4 (solid line) and 1 M KOH (broken line) [146]. (B) Δ versus ψ relationship for anodic oxide formation (solid line) and reduction (broken line) in 0.5 M H_2SO_4 [146]. (C) Same as (B) but for 1 M KOH [146].

interpreted by Sirohi and Genshaw [145] as the manifestation of the chemisorption of oxygen in the form AuOH or AuO. The ψ-Δ plots indicate an identical behavior in acidic, alkaline, 1 M KF, and K_2SO_4 solutions; this rules out the anion effect since the same phenomenon is observed for different anions. However, a clearcut explanation for the observed behavior cannot yet be given. From the coulometric data and the assumption of a monolayer thickness of 3 Å, Sirohi and Genshaw [145] find the film optical constant to be 0.3-0.7i (n - ki). These values correspond more to a metallic surface than for an adsorbed species. On the other hand, Vinnikov et al. [147] find a range of 3 to 4 for the refractive index n, and 1.0 to 1.5 for the absorption coefficient k. The exact value of these constants can only be determined if the film thickness is known. In their case, Vinnikov et al. [147] assume a roughness factor of 5-6 which seems unreasonably high. They conclude that, to explain their data the film must consist of a mixture of electrostatically-attracted species and discharged particles. Sirohi and Genshaw [145] conclude that the absorbed species influences the optical properties of the metal only.

For potentials in the so-called phase-oxide region, i.e., around 1.4-1.45 V (versus RHE), new values of the optical constants for the film for a monolayer of oxide 5.5 Å thick were estimated to be 2.7-0.28i [147]. Such an evaluation was not carried out by Sirohi and Genshaw [145]. However, the latter authors tried to correlate the changes in Δ with the time, as the potential was stepped from 1.24 V to E_a (where E_a ranged from 1.34 to 1.84 V) in 100-mV increments. Neither the direct logarithmic (Δ versus log t) nor the inverse ($1/\Delta$ versus log t) rate law can account for the experimental data. This gives rise to some doubts about the ability of the optical parameters to follow the kinetics of the oxide formation. The process is too complex to permit the use of only the optical parameters to follow the kinetics of the gold oxide (Au_2O_3 or its hydrated form) formation.

Reflectance studies on the gold electrode yield some interesting data [99,100,106,107,150,151]. Figure 29 shows the specular

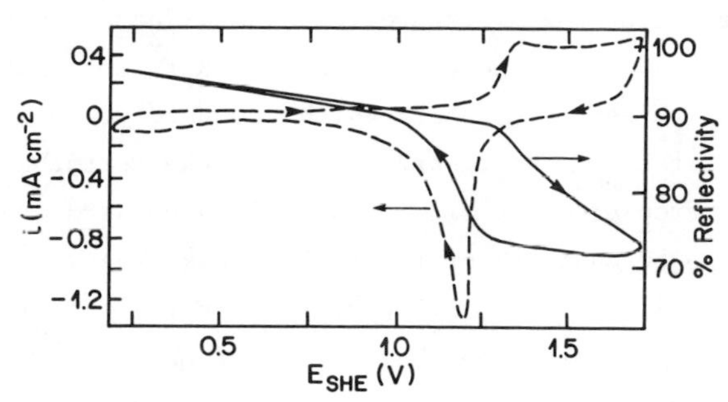

FIG. 29. Simultaneous cyclicvoltammogram (broken line) and specular reflection curves (solid line) for a gold electrode in 0.2 M $HClO_4$ at 22°C; v = 108 mV sec^{-1}, 5,400 Å, angle of incidence 50°, 19 reflections [150].

reflection versus potential data as well as the current-potential cyclicvoltammetric curve of Takamura et al. [150]. The reflectance data represent multiple reflections (∿20 reflections); these data are qualitatively similar to the ones obtained for a single reflection [99]. We note a slow linear decrease in reflectivity from 0.24 to 1.29 V (versus SHE); as this potential is exceeded, a rapid decrease in reflectivity occurs with a net change of slope at approximately 1.45 V. This is the potential range where the oxide film is being formed. The break at 1.45 V seems to correspond to a monolayer of oxide (AuO).

The change in reflectivity with incident wavelength can also yield important information as to the nature of the gold surface. The plot of the function $1/R\ (dR/dE)_\tau$ against the wavelength for different potentials gives rise to a maximum at approximately 5000 Å and this maximum undergoes a bathochromic shift (shift toward longer wavelength) as the potential of the electrode increases. The maximum is attributed to the electronic transition in gold (5d → 6s) at approximately 5000 Å so that below this wavelength light absorption occurs that decreases the reflectivity. However, this interpretation was contested by McIntyre and Kolb [106], who maintain that the data are best represented by a charge-transfer absorption in the semiconducting oxide film.

The origin of the electroreflectance effect in metals is a controversial subject (see discussions in Refs. 92 and 93) and as yet there is no definite theory which adequately explains all the experimental facts. The theory involving the perturbation of the tail of the free electron plasma tunnelling outward from the surface fails to account, according to Lazorenko-Manevich and Stoyanovskaya [107], for the large variation in the relative changes with potentials observed on the metal surface in the absence of oxides. However, these authors [107] do not propose an alternative theory to account for their experimental results. One difficulty involved in the assessment of these studies is the differences in the experimental techniques (optical and electrochemical) and the electrode pretreatments used by various workers.

Kim et al. [152] confirmed by ESCA the presence of Au_2O_3, or more precisely, its hydrated form, $Au(OH)_3$, produced by the anodic oxidation of a gold electrode at 1.8 V and 1.9 V in 0.5 M H_2SO_4.

The anodic oxidation and reduction of gold has also been studied using a new technique based on the measurement of the coefficient of friction of an electrode undergoing potential scanning; it is called polaromicrotribometry [153-155]. Figure 30 depicts the current versus potential and the coefficient of friction versus potential results for a gold electrode in 0.5 M H_2SO_4. The coefficient of friction versus potential curve has exactly the same characteristics as the ellipsometric data [145]. It shows the well-pronounced hysteresis between the reduction and the oxidation of the gold electrode.

C. Conclusions

Even though numerous studies have been carried out to elucidate the oxide formation on gold, some points still remain unclear.

The oxidation at high anodic potentials [greater than 2 V (versus RHE)] leads to the formation of a phase oxide of the composition Au_2O_3 or its hydrated counterpart. Upon reduction of this deep red oxide, the gold electrode maintains its dark color passing

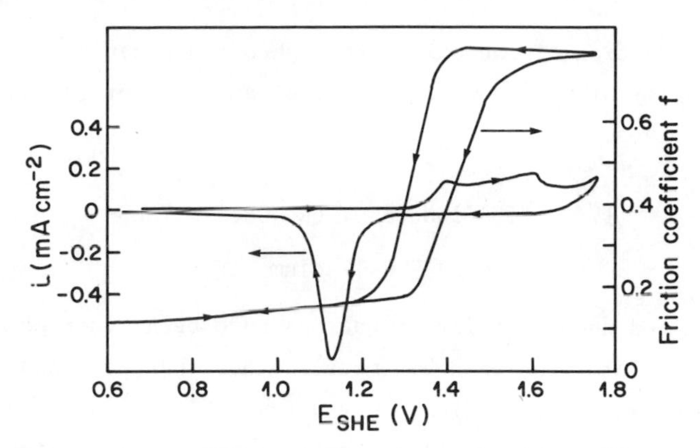

FIG. 30. Simultaneous cyclicvoltammogram and polaromicrotribometric curve for a gold electrode in 0.5 M H_2SO_4; v = 17 mV sec^{-1}, load on the friction measuring rod = 5 g, with a displacement speed of 8 μm sec^{-1} [154].

from red to black. This occurs because of the formation of finely-divided gold particles, thus giving rise to a gold-black (similar to a platinum-black) surface. However, below this high anodic potential limit, the surface undergoes gradual oxidation as shown by ellipsometric and electroreflectance data. A monolayer of oxide of the type AuO is obtained at approximately 1.45 V (versus RHE) in an acidic medium. From galvanostatic charging or cyclicvoltammetric curves, the onset of the oxide formation is around 1.3 V. In the latter type of experiments, several peaks are observed in the oxide region, indicating the possibility of different types or structures of oxides, as in the case of platinum.

Time effects are also observed; the longer the time of oxidation, the higher the overpotential needed for the reduction of the formed oxide. Under some particular experimental conditions (low temperatures or fast scanning speed), the reduction peak can be resolved into as many as three components.

Finally, one problem encountered in the study of the oxidation of gold is its anodic dissolution as well as the chemical dissolution of the electrochemically-formed oxide. This dissolution has to be taken into account in the evaluation of the charges associated with the reduction of the oxide; this phenomenon can account for the inequality of Q_a and Q_c in cyclicvoltammetric or galvanostatic charging curves.

IV. OXIDATION OF OTHER NOBLE METALS

A. Palladium

Only a limited number of investigations have been reported on the anodic oxide formation on noble metals other than platinum and gold.

In the case of palladium, only a few publications have appeared since the review of Hoare [1]. A short summary of the studies carried out prior to 1965 is given below together with a detailed discussion of more recent papers.

There is evidence from the charging curves [156,157] that two types of oxides are formed on palladium, namely PdO and PdO_2. The anodic charging curves exhibit arrests at anodic potentials which agree with the thermodynamically-predicted potentials for the formation of these species. These arrests have not been detected by some other workers [158] who attributed the discrepancy to the presence of impurities which gave spurious arrests in the previous work [156,157]. However, it is well established that the presence of an oxide or chemisorbed oxygen is indicated at approximately 0.8 V (versus RHE) in the charging curves [156-158] or the potentiodynamic profiles [159]. On the basis of the measurement of rest potential of chemically-prepared PdO, the arrest observed in the cathodic charging curve of the previously-anodized palladium is indicated to be associated with the reduction of PdO [157].

On a reduced electrode, the palladium rest potential in oxygen-saturated solutions is approximately 870 mV (versus RHE) [160]. This potential corresponds to the Pd/PdO electrode and pertains to the following equilibrium:

$$PdO + 2H^+ + 2e \rightleftarrows Pd + H_2O \qquad (55)$$

The experimental rest potential corresponding to the above equilibrium was found to depend on the oxygen partial pressure [160].

To account for such dependence, the observed rest potential was postulated to be a mixed potential arising from the following reactions:

$$Pd \rightarrow Pd^{2+} + 2e \quad (56)$$

$$O_2 + Pd^{2+} + 2e \rightarrow PdO_2 \quad (57)$$

The dependence of the rest potential on the oxygen pressure was also observed by Schuldiner and Roe [161].

In recent years a great deal of effort has been devoted to the examination of the corrosion of palladium in acidic media. Cadle [162], using a ring-disk electrode, found the corrosion rate of palladium polarized at 1.34 V (versus SHE) to be 0.35 $\mu A\ cm^{-2}$. A similar value was also reported by Llopis et al. [163] using a radiotracer technique. In an experiment similar to that performed on gold [135], the ring current was recorded when the disk potential was scanned from 0.24 to 1.64 V (versus SHE) while the ring potential was held constant at 0.24 V. The result is illustrated in Figure 31. The cyclicvoltammetric curve on palladium is similar to those on other noble metals in that a hysteresis is observed between the oxide formation and its reduction. A deposition current is observed on the ring as the disk potential exceeds 0.7 V (versus SHE); also a peak is observed as the reduction of the palladium oxide takes place. For potentials above 1.37 V, oxygen reduction is observed. It was shown that the species deposited during the potential scan is Pd(II). The corrosion current decreases as a function of time as the palladium electrode undergoes passivation. Also, if the potential is held at anodic values greater than 1.3 V, another soluble species is detected during the reduction of the palladium oxide formed at this high potential. This was attributed [162] to the reduction of a phase oxide formed at the high anodic potentials. The phase oxide formation was also noted by Rand and Woods [139] but no limiting oxide coverage was observed, in contrast to the case of platinum or rhodium (see Section IV.B). Palladium behaves like gold in this respect; a small limiting oxide coverage is achieved in the

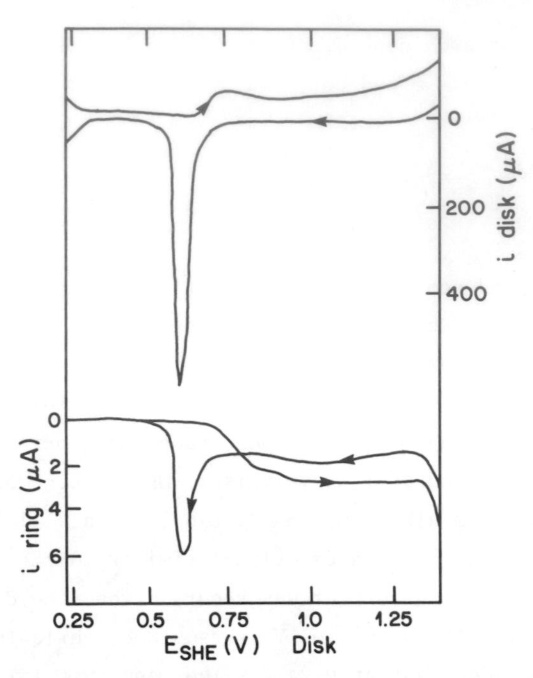

FIG. 31. Rotating ring-disk palladium electrode; disk cyclic-voltammogram in 0.2 M H_2SO_4, v = 100 mV sec^{-1}, ω = 2,500 rpm; constant ring potential held at 0.24 V [162].

potential range of 1.5 to 1.7 V (versus RHE), and beyond this potential, the oxide coverage increases and also a roughening and coloring of the surface occurs. The stoichiometry of the oxide at potentials around 1.5 V is concluded to be one oxygen atom per palladium atom, i.e., the PdO species. This species was detected and measured by X-ray photoelectron spectroscopy (ESCA) by Kim et al. [164] who compared the electrochemically-prepared palladium oxides with samples prepared by heating palladium in air at different temperatures.

Kim et al. [164] could identify the presence of adsorbed oxygen as well as the PdO in the chemically-prepared samples. For polarizations of palladium below 0.8 V (versus SHE), in 0.5 M H_2SO_4, no phase oxide (PdO) could be detected. However, for samples polarized at 0.9 V (versus SHE), two oxide peaks were observed, one clearly related to the PdO oxide and the other corresponding to PdO_2. The latter species could not be prepared chemically so that its assignment is somewhat speculative and rests on arguments similar to the ones used in the interpretation of the spectrum of oxidized platinum. The presence of the PdO_2 species concluded in this work [164] confirms the mechanism proposed by Hoare [160] to account for the rest potential of 0.87 V observed in oxygen-saturated acidic solutions [see Eqs. (55)-(57)]. Also, the ESCA spectrum shows that the oxide is formed uniformly on the surface, and for samples polarized at 0.90 V for 1,000 sec, Kim et al. [164] estimate the coverage to be 2-3 oxide layers. At higher polarizations [1.7 V (versus (SHE) for 30 sec] the palladium metal peak completely vanishes and only the PdO and PdO_2 peaks are observed; for this surface an estimated oxide thickness of 40 Å is proposed. These ESCA experiments clearly show the presence of an oxide (PdO) above 0.8 V (versus SHE) in acidic solutions; at potentials above 0.9 V the presence of a PdO_2 oxide is indicated. The PdO_2 presumably originates by a mixed potential mechanism below its formation potential. At higher potentials this type of oxide could be the predominant species.

Tarasevich et al. [165] examined the cyclicvoltammetric behavior of theoxide reduction of palladium both in acidic and alkaline solutions. In an alkaline medium, they observed two peaks in the anodic scan, the first at 0.7 V (versus RHE) being reversible and the second at higher anodic potentials exhibiting the familiar hysteresis between the oxide formation and reduction. From the variation of the peak potentials with sweep rate, they propose a mechanism involving interactions between the deposited (M-O or M-OH) species. In this analysis, the following reaction mechanism is suggested:

$$H_2O + Pd \rightarrow PdOH + H^+ + e \tag{58}$$

$$2PdOH \rightarrow PdO + Pd + H_2O \tag{59}$$

The reduction is proposed to involve the following pathway in alkaline solutions:

$$PdO + H_2O + e \rightarrow PdOH + OH^- \tag{60}$$

$$2PdOH \rightarrow PdO + Pd + H_2O$$

or

$$PdOH + e \rightarrow Pd + OH^- \tag{61}$$

In acidic solutions, the suggested reactions are:

$$PdO + H^+ + e \rightarrow PdOH \tag{62}$$

$$2PdOH \rightarrow PdO + Pd + H_2O \tag{59}$$

At low sweep rates, for reactions (62) and (59) with (59) in equilibrium, a rate law is derived in which the experimental variation of the potential maximum (of the potentiodynamic peak) with the sweep rate is deduced. As the sweep rate increases, reaction (59) becomes rate-determining and the experimental dE/log v of 120 mV is obtained for such conditions.

Capon and Parsons [130] did not observe a large concentration effect on the cyclicvoltammetric curves for different sulfuric acid concentrations from 1 to 7.5 M. A slight shift in the potential of the anodic oxide formation and reduction peaks is observed at the highest concentration used indicating possible effects of anions or the effect of decreasing water concentration as the acid concentration increases.

B. Rhodium

Shibata [166] has examined the oxide formation and reduction on rhodium electrodes by means of galvanostatic transients. At constant current, a linear increase of potential with time from 0.6 V (versus SHE) to almost 1.6 V (versus SHE) was observed. In this potential region oxidation of the metal takes place. In conjunction with charging curves, Shibata [166] measured the electrode capacity and observed three main types of behavior. At the onset of the oxide formation, the capacity decreases; from 0.9 V to 1.2 V the capacity remains constant and finally increases to a constant value above 1.5 V when the oxygen evolution reaction commences at detectable rates. These observations were explained as follows. First, there is the discharge of water to form the RhOH species; this formation has some reversible character. As the potential increases the RhOH is further oxidized to RhO and finally to the stable Rh_2O_3 oxide or its hydrated form RhOOH. It was found that the charge needed to oxidize the electrode exceeded the charge consumed during the reduction process. This was also concluded by Böld and Breiter [167] from the potentiodynamic curves. However, this inequality was attributed to the dissolution of the electrode [24], as for other noble metals (platinum, gold, and palladium). By considering the dissolution of the metal as well as by paying particular attention to the impurity problem, Rand and Woods [24] could obtain a quantitative agreement between the anodic and cathodic charges. These authors also studied the oxide coverage as a function of anodic polarization [139] and observed a limiting coverage as for the case of platinum. Their results are shown in Figure 32. A small arrest corresponding to a value of $Q_O/2Q_H = 1.7$ may be noted. Considering the difficulty in assessing the value of the charge associated with

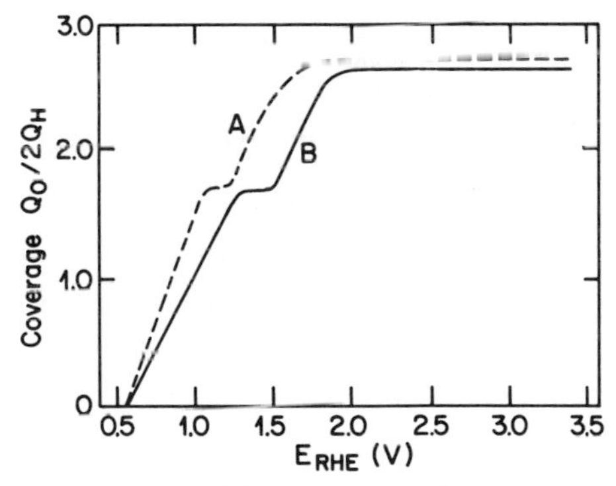

FIG. 32. Oxide coverage as a function of potential for a rhodium electrode in 1 M H_2SO_4 at 25°C; anodic polarization for 1,000 sec (A) and 10 sec (B) [139].

full coverage by hydrogen (Q_H), in relation to the real surface of the electrode, they approximate this arrest to a one-to-one oxygen-rhodium species (i.e., a monolayer of Rh-O). The limiting oxide coverage indicates high electronic conduction of the oxide permitting the oxygen evolution reaction to take place easily. The same comment applies to platinum. Poor electronic conduction through the oxide will lead to thick oxide formation as in the case of palladium and gold [168].

In the potentiodynamic profiles, an anodic shift in the oxide formation peak, with increasing concentration of the electrolyte, is observed as in the case of palladium [130]. Also, Capon and Parsons [130] noted a second hydrogen oxidation peak as the sulfuric acid concentration was increased from 1 to 7.5 M.

Hoare [169] reported on the rest potentials of the rhodium electrodes in oxygen-saturated solutions. The Rh/RhO system has a mixed rest potential of 0.93 V (versus SHE) and the Rh/Rh_2O_3 system exhibits a mixed rest potential of 0.88 V (versus SHE).

In a series of investigations on the oxidation of noble metals, Burshtein et al. [170] also examined the rhodium electrode. They proposed the following mechanism for the oxidation of Rh:

$$H_2O + Rh \rightarrow RhOH + H^+ + e \tag{63}$$

$$RhOH \rightarrow RhO + H^+ + e \tag{64}$$

At low sweep rates, reaction (63) is quasireversible and (64) is irreversible in the anodic direction of the scan; the reversibility of reactions (63) and (64), respectively, are reversed during the cathodic sweep. As the sweep rate is increased to above 10 V sec^{-1}, both reactions behave irreversibly. From this proposed mechanism, Burshtein et al. [170] derived expressions relating the potential of the current maximum to the logarithm of the sweep rate. At low sweep rates, the shifts in the peak potentials are 30 mV per decade of the sweep rate with the reversible conditions assumed for reaction (63). As the sweep rate increases, the derived peak potential shifts become 120 mV per decade and are comparable to those observed experimentally, viz. 130-140 mV per decade.

In another paper [171], Khrushcheva et al. studied the time effects in the cyclicvoltammetric oxide formation and reduction. For these experiments, a trapezoidal voltage scanning program was used. If the anodic potential, in acidic media, is held at 1.0 V (versus RHE), the reduction peak of the so-formed oxide is shifted by a few millivolts in the cathodic direction. However, if the anodic oxidation potential is 1.4 V, the reduction peak increases in magnitude as the polarization time increases and the cathodic shifts become much more important. From these results and the similar ones obtained in alkaline solutions, Khrushcheva et al. [171] concluded that there were two types of oxides on the rhodium surface and that the onset of the formation of the most stable oxide occurred at 1.4 V. The first form is interpreted as chemisorbed oxygen (up to 1.0 V) of the type Rh-O, as also proposed by Hoare [169]. As the potential increases, the oxide Rh_2O_3 begins to be formed and since the overpotential of its reduction is higher than that for the chemisorbed RhO oxide (the Rh_2O_3 seems to be the stable form of the oxide species). The presence of two types of oxide was also reported by Llopis and Vazquez [172] from the galvanostatic reduction of pre-anodized rhodium electrodes. However, their oxide reduction curve differs from that obtained by Shibata [166] where only one arrest was observed at 0.4 V. Llopis and Vazquez [172] observed a second arrest at 0.8 V in concentrated (2 M) $HClO_4$; the presence of some residual oxygen gas could possibly account for this second arrest in the reduction curves and their results have thus to be taken with some caution.

C. Iridium

Amonst the noble metals iridium has some unique characteristics as regards its oxidation behavior; these features, together with other related topics, are outlined below.

On potentiodynamic cycling, the electrode oxidation shows the characteristics of a reversible reaction. Figure 33 depicts the potentiodynamic potential-current relationships in acidic and

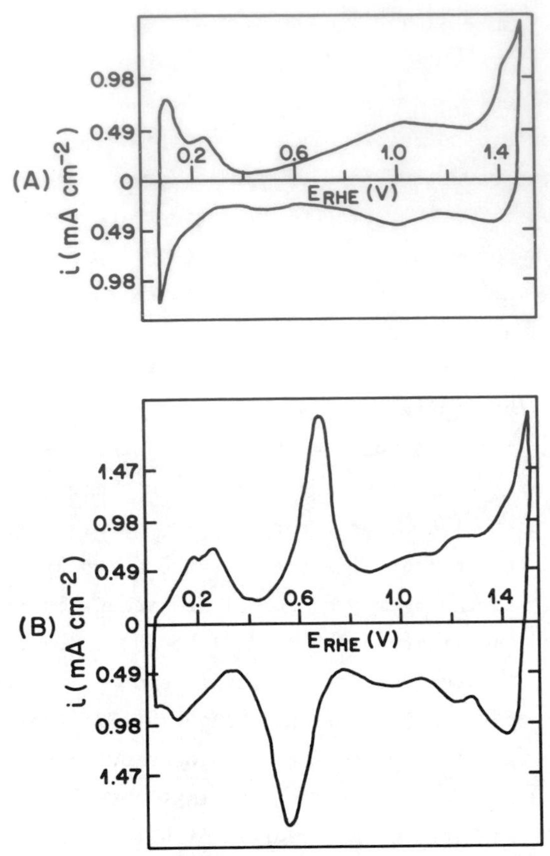

FIG. 33. Cyclicvoltammograms of an iridium electrode in 2.3 M H_2SO_4 (A) and 0.1 M NaOH (B); v = 0.5 V sec^{-1} at 25°C [167].

alkaline media for the oxide formation and reduction [173]. These curves are consistent with those reported by several other authors [130,174-177]. The reversible character of the oxide formation is illustrated in the instantaneous change of sign of the current as the potential scan is changed from the anodic to cathodic direction [174]. Also, the anodic peak and the corresponding cathodic peaks do not exhibit the pronounced hysteresis characteristic of other noble metals. The reversible character of the oxide formation was confirmed by the optical measurements of Conway and Gottesfeld [104].

They showed that the differential reflectance $\rho(V) = (\Delta R/R)(1/\Delta V)$ was completely reversible as the potential was scanned from 0.2 to 1.4 V (versus RHE) and, in contrast to platinum, it did not exhibit irreversibility when the electrode potential was held at 1.3 V for 120 sec; for iridium the $\rho(V)$ values in the cathodic scan, after holding the potential at the anodic limit, are the same as those obtained for a continuous potential cycle. Conway and Gottesfeld [104] concluded that the following reactions were reversible:

$$Ir + H_2O \rightarrow IrOH + H^+ + e \tag{65}$$

$$IrOH \rightarrow IrO + H^+ + e \tag{66}$$

The behavior of iridium on repeated potentiodynamic cycling is unique as reported by Capon and Parsons [130] and examined more extensively by Otten and Visscher [177] and Rand and Woods [176]. As the number of cycles increases, the anodic oxide formation current as well as its reduction current increase continuously. This effect is not due to a surface roughening since the hydrogen adsorp- and oxidation peaks are not modified to such an important extent [177]. An ellipsometric study of this effect was also carried out by Otten and Visscher [177] who found a linear relationship between the ellipsometric parameters $\Delta\Delta°$ and $\Delta\psi°$ and the number of potentiodynamic cycles. This linear relationship was also observed between the oxide charge (Q_O) and the number of cycles; the charge Q_O was measured by integrating the oxide reduction current. This effect is interpreted as a gradual formation of a "disturbed" metal layer. The thickness of this layer can be estimated from ellipsometric data to range from 15 to 49 Å after 100 cycles for different reasonable values of n and k. The thickness of the oxide layer, estimated from the oxidation charge Q_O after 120 cycles is only 3.3 Å. The discrepancy is accounted for by assuming a pit oxidation model where the oxide is formed preferentially in pits or crevices of the iridium electrode. The nonuniform thickness of the oxide layer was also observed in field-ion microscopic studies [117].

The oxide coverage on iridium electrodes that had undergone different pretreatments has been examined by Otten and Visscher [178]. Their main result was the resolution of the two types of oxides. The oxide coverage versus potential curve shows a break at potentials around 1.1 V (versus RHE). The same break was observed in their ellipsometric parameter versus potential curves. These data fall in line with the previous determination of the oxide coverage on iridium by Damjanovich et al. [179] where such a break or change of slope was also observed. These data were interpreted as an indication of the formation of different types of oxides [178]. The mechanism for the oxide formation, taking into account the pit model as discussed above, is proposed [178] as:

(a) Oxide formation in the pits below 1.2 V
(b) Pits completely full of oxide product at 1.2 V
(c) Above 1.2 V, metal layer further attacked and the penetration becomes more important

Recently several studies on the electrochemical behavior of iridium have also been published by the Soviet workers [175,180-184]. The final conclusion reached by Kurnikov and Vasil'ev [184] is that iridium oxides are formed via the IrOH, IrO_2, and finally the Ir_2O_3 species in acidic solutions. The formation of these species can be followed from the potentiodynamic curves in which different peaks associated with these species appear. The detailed analysis of the variation of the peak maxima with the potential sweep rate and pH seems to confirm their proposed mechanism. It must be noted that the oxide formation on iridium is a slow process and at fast scanning rates ($v > 5\text{-}10\ V\ sec^{-1}$) oxygen diffusion into the oxide becomes a limiting factor. The involvement of the IrOH species agrees well with the optical data obtained by Conway and Gottesfeld [104].

The ESCA spectrum of Kim et al. [152] confirmed the presence of IrO_2 for anodic polarizations from 1.16 to 1.40 V. An oxide thickness of 55 Å was estimated from the ESCA spectrum and electrochemical measurements, for an anodization of 30 sec at 1.40 V.

The increase in anodic and cathodic current upon repetitive potential cycling was also studied by Rand and Woods [176]. Such an effect was observed for anodic limits exceeding 1.4 V. The process for such behavior is postulated to be a buildup of a phase oxide with a change of oxidation state. Upon cycling, the oxide layer is not completely reduced and so the higher oxidation state can be achieved. The oxide so formed can be removed from the electrode surface by cycling the electrode in more concentrated acidic solutions (5 M). The change of oxidation state of an irreversibly-formed phase oxide is claimed by Rand and Woods [176] to confer the reversible character to the cyclicvoltammogram; the reversible features observed by several authors [130,174-177] are attributed to the change in the oxidation state of the phase oxide already present on the electrode and not to the reversible chemisorption of [O] or [OH]. However, this point will be debated for some time before any definite conclusion is reached.

D. Ruthenium

There are very few recent studies of the ruthenium oxide formation or dissolution. Under anodic polarization in acidic or alkaline solutions, ruthenium undergoes dissolution. These processes have been studied by Llopis et al. [185,186] and by Gorodetskii et al. [187]. Trasatti et al. [188] and Galizzioli et al. [189] have reported on the electrochemical behavior of chemically-prepared RuO_2 and they compared it with the behavior of electrochemically-grown RuO_2 on a ruthenium electrode. The cyclicvoltammetric curves for these two electrodes are presented in Figure 34. From these curves they concluded that the formation of RuO_2 on ruthenium occurs at approximately 1.0 V, following the onset of the metal oxidation at 0.55 V. On ruthenium electrodes, no definite separation between the hydrogen adsorption and the oxide reduction regions is observed. Recently, Hadzi-Jordanov et al. [190] were able to show more clearly the hydrogen adsorption peaks on electroplated ruthenium. This potential range is masked by the oxide reduction if the potential sweep

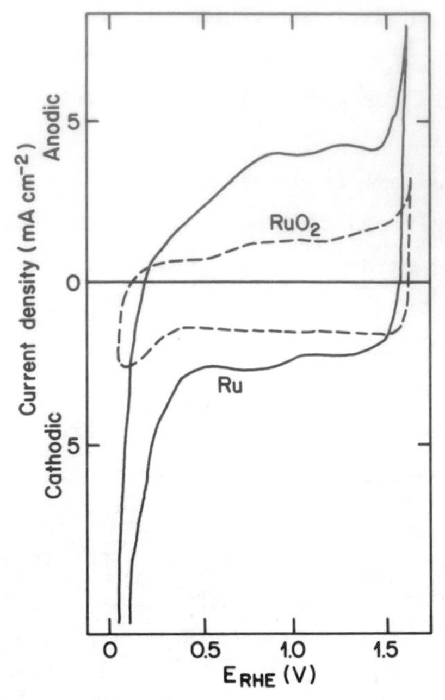

FIG. 34. Cyclicvoltammograms of a ruthenium electrode and a deposited RuO_2 film; v = 40 mM sec^{-1} in 1 M $HClO_4$ [183].

is extended beyond 0.5 V. As the potential scan rate increases, the cyclicvoltammetric curve becomes more structureless [191]. Similar studies were also carried out by O'Grady et al. [192]. One interesting feature of this metal is its low overvoltage for the oxygen evolution reaction [183]. An identical potential-current behavior is observed for the ruthenium and the RuO_2 electrodes. However, with the former, the solution acquires a yellow color during prolonged oxygen evolution. This behavior related to the anodic dissolution of ruthenium into soluble species (RuO_4 or H_2RuO_5) is not observed for the RuO_2 electrode. Similar electrochemical characteristics were also reported by Burke and O'Meara [193] for the electrodeposited ruthenium on platinum.

E. Osmium

Of all the noble metals, osmium is the least examined. Charging curves for the electrodeposited osmium electrodes were obtained by Khomchenko et al. [194] and by Appleby [195]. From these curves they [194,195] could observe the hydrogen desorption, the double layer and the oxygen adsorption regions. The charging curves are similar to those for the ruthenium electrodes. The potential-charge curves derived from the anodic and cathodic charging curves are shown in Figure 35. The hydrogen desorption region extends up to approximately 150 mV; a smooth transition to the oxide formation is then observed. This oxide formation potential is not very well defined and its exact value is difficult to assess from these charging curves. Above 0.88 V the osmium dissolution reaction is observed. A small hysteresis in the charging curves is observed as for most noble metals. This hysteresis is not as important as that observed on platinum. Cyclicvoltammetric curves on osmium were reported by Llopis and Vazquez [196] and are similar to the ones obtained for ruthenium. No detailed mechanisms for the oxide formation or the dissolution of osmium were proposed.

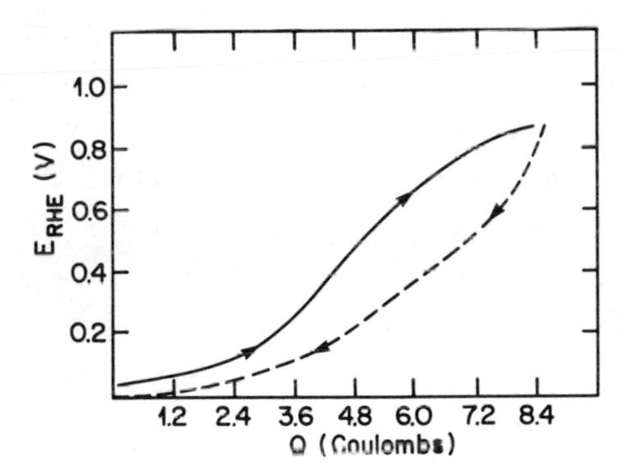

FIG. 35. Potential as a function of charge for an osmium electrode deposited on platinum electrode in 0.05 M H_2SO_4; solid line, anodically increasing potentials; broken line, cathodically increasing potentials [194].

V. OXIDATION OF NOBLE METAL ALLOYS

A. Introduction

The study of noble metal alloys is an important research tool in that the d-band character of the metal, and thence its electronic properties, and the lattice parameters can be systematically varied in order to seek their possible correlation with the catalytic, electrocatalytic, oxidative, and other surface properties. An important point to bear in mind in such studies is the exact nature of the alloy (homogeneous dissolution of the components, phase formation) and, in particular, the exact composition of the alloy surface. The importance of these factors in the anodic oxide growth on noble metal alloys will be pointed out, where pertinent, in the following discussion.

B. Platinum-Gold

Platinum and gold form a series of solid solutions at high temperatures (temperatures just below the solidus [197]) but as the temperature is lowered two phases develop: the α_1 phase is platinum-rich and the α_2 phase is gold-rich [197-199].

A typical cyclicvoltammetric scan for a platinum-gold alloy electrode is illustrated in Figure 36 together with the potentiodynamic current-potential relationship for the respective pure metals [197]. The cyclicvoltammogram for the alloy has the characteristics of the constituent metals. The metals in the alloy behave independently from one another and there is no indication from the work of several authors [197-199] that a peak specific to a platinum-gold alloy arises. However, Michri et al. [200] detected an oxide reduction peak located in between those for the reduction of oxides of gold and platinum; this new peak is more pronounced for high gold content alloys, namely, those containing 70 and 90% gold. This third peak indicates a new surface phase but its exact identity could not be established by X-ray analysis. New analytical tools for the surface analysis (ESCA and Auger spectroscopy) could be useful for establishing the exact composition of the surfaces of these alloys.

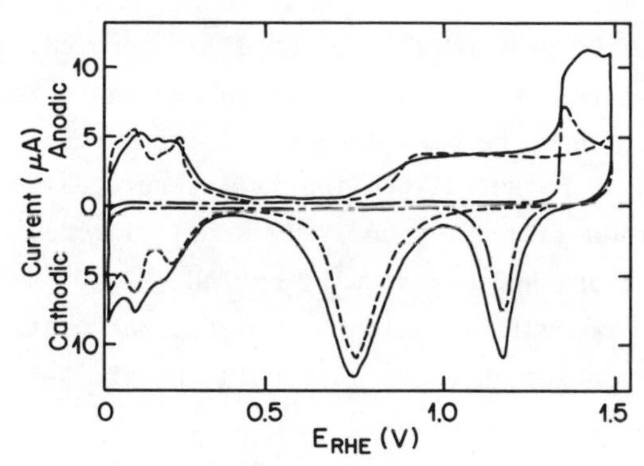

FIG. 36. Cyclicvoltammogram of a 65% platinum, platinum-gold alloy electrode (—) with, for comparison purposes, the curves for pure gold (— - —) and pure platinum (- - -) electrodes; v = 40 mV sec^{-1}, 1 M H_2SO_4 [197].

The repetitive cycling of the electrode does not alter significantly the overall characteristics of the curves for the platinum-gold alloys.

From the additivity of the cyclivoltammetric curves, Woods [197] concluded that the electrode surface consists of two phases (the α_1 and α_2 phases, being platinum- and gold-rich respectively) and each phase behaves independently. Woods [197] could not detect the intermediate peak observed by Michri et al. [200].

The absence of any new features for the platinum-gold system is also confirmed by studies on electrodeposited platinum-gold alloys [201]. In this case the cyclicvoltammogram exhibits the oxide reduction peak of the individual metals. It can thus be concluded that for homogeneous or electrodeposited alloys, the platinum-gold system behaves electrochemically as an agglomerate of gold and platinum, with no change in the electrochemical properties of the component metals brought about by alloying.

The surface of the alloy is composed of definite platinum and gold aggregates with a minimum interaction between one another. Woods [202], using X-ray diffraction profiles, studied the surface (several angstroms deep) composition of platinum-gold alloys as a function of potential cycling. On an untreated electrode (an electrodeposited platinum-gold electrode containing 65% platinum) a single diffraction peak related to gold is observed. As the potential cycling is carried out, a second peak associated with the pure platinum metal becomes apparent. The surface composition at the beginning of potential cycling is interpreted as being composed of a platinum-gold alloy and very small platinum crystallites (less than 50 Å); as the electrode is cycled (27,000 potential cycles applied), recrystallization occurs and the platinum crystallite size increases enough to be detectable by the X-ray diffraction technique.

C. Platinum-Rhodium

The oxide formation and reduction on platinum-rhodium electrodes has been studied rather intensively [203-206]; a typical cyclicvoltammogram for a 26% platinum, platinum-rhodium is illustrated in Figure 37. In the figure the first and the 1,000th sweep are illustrated. On the first sweep a single oxide reduction peak is observed and the potential of the peak maximum is intermediate between the value of the pure metals constituting the alloy (the oxide reduction peak maxima for the pure metals are also shown in Figure 37). In fact a linear relationship can be established between the potential of this peak maximum and the platinum content

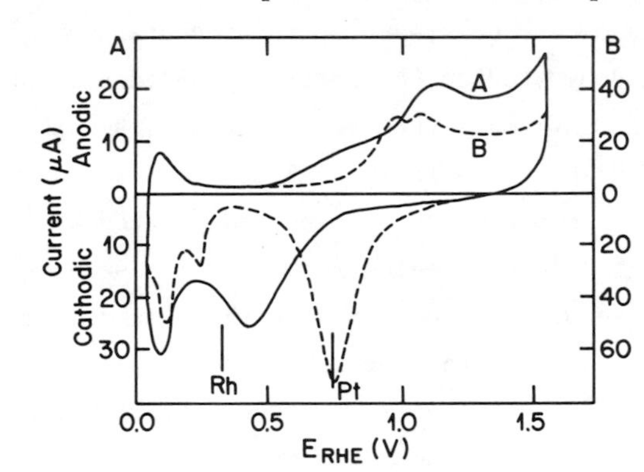

FIG. 37. Cyclicvoltammograms of 26% platinum, platinum-rhodium alloy; 1st cycle (A) and 1,000th cycle (B); at 25°C, v = 40 mV sec^{-1} in 1 M H_2SO_4. Also indicated are the peak positions for the oxide reduction of the pure metals under identical experimental conditions [203].

of the alloy [203,205,206]. As the number of potential sweeps increases, this aspect of the current-potential curve is greatly modified. The oxide reduction peak is shifted to more anodic potential values and finally the electrode behaves as a pure platinum electrode [203-206]. This behavior is associated with the different rates of dissolution of the alloy components. Rhodium will dissolve more rapidly and the final surface composition will be composed of the metal undergoing the slowest dissolution (in the present case, platinum). The dissolution rates were measured [203] and it was found that rhodium was dissolving approximately six times faster than platinum. It is important to note that this surface modification could not be detected by X-ray diffraction studies. The uncycled and cycled electrode spectra being identical [137]. The electron probe microanalysis was found to be not sensitive enough to detect this surface change.

The linear relationship between the surface composition and the oxide reduction peak can thus be a very sensitive tool for establishing the surface composition of an alloy.

For example [203], the 26% platinum, platinum-rhodium alloy after 20 cycles had an effective surface concentration of platinum of 50%; the oxide reduction peak was observed at 0.55 V.

Heat treatment of the alloy modifies its electrochemical behavior; the oxide reduction peak is shifted to more cathodic values after heat treatment (heating to redness in a natural gas-air flame) indicating a surface enrichment by the rhodium [203].

The different behavior of physically-mixed platinum and rhodium black as well as codeposited platinum-rhodium electrodes has been studied by the cyclicvoltammetric method [204]. For the physically-mixed electrode, the current-potential curves indicate the additive behavior of platinum and rhodium. Upon repetitive cycling, the platinum oxide reduction peak decreases in intensity due to the dissolution and redeposition of rhodium in the porous matrix electrode used. The X-ray diffraction pattern showed the presence of discrete platinum and rhodium metal areas for such electrodes. For codeposited platinum-rhodium electrodes, the cyclicvoltammetric curves as well as the X-ray diffraction pattern indicate the presence of an homogeneous alloy; the behavior of such an electrode was similar to that of a smooth platinum-rhodium alloy. The same conclusion (i.e., alloy formation of codeposited platinum-rhodium electrode) was reached by Rand and Woods [201]. The broadness of the oxide reduction peak for such an electrode indicates a range of composition of the so-formed alloy.

D. Gold-Palladium

The gold-palladium system has been investigated by several workers. The work on the electrochemical behavior of the smooth alloy has been reported by Rand and Woods [203]. They also examined deposits of palladium on gold [201]; this system was also studied by Cadle [207]. Finally, Woods [208] investigated the codeposits of these metals on a tantalum foil. Figure 38 shows the cyclicvoltammetric curve of a smooth Pd-24% Au alloy as a function of the number of potential cycles. The oxide formation and, especially, the oxide reduction peaks do not coincide with either pure gold or pure palladium. The oxide reduction peak potential is a linear function of the alloy composition; similar behavior for the platinum-rhodium alloy system has been noted above. However, as the number of potential cycles increases, the oxide reduction peak at 0.81 V (for a 76% palladium, palladium-gold alloy) decreases in intensity and a new peak which is associated with the pure gold electrode appears at 1.20 V. The cyclicvoltammogram after 1,000 cycles is more or less identical with that of a pure gold electrode. This behavior for the palladium-gold system is different from that observed for platinum-rhodium; in the former case, the oxide reduction peak potential does not vary as a function of the number of cycles, indicating that the surface composition does not change. The removal of palladium from the alloy surface does not result in a continuously-varying surface composition of the alloy.

For a codeposited palladium-gold electrode a similar effect is observed [208]. X-ray analysis indicates that palladium is removed

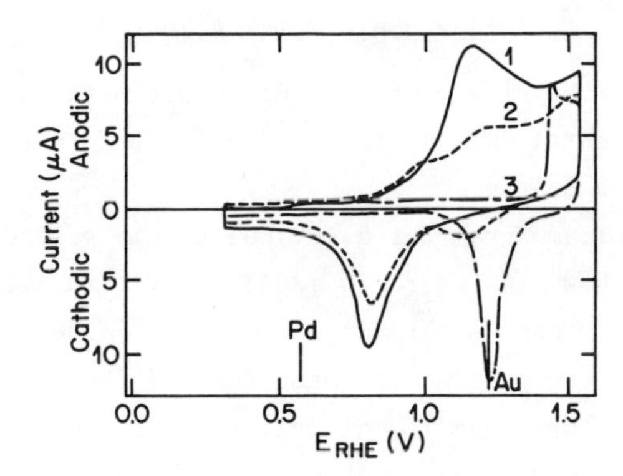

FIG. 38. Cyclicvoltammograms of a 76% palladium, palladium-gold alloy; 1st (1), 3rd (2), and 1,000th (3) cycles; at 25°C, v = 40 mV sec^{-1} in 1 M H_2SO_4. Also indicated are the peak positions for the oxide reduction of the pure metals under identical experimental conditions [203].

from the surface but not from the bulk of the codeposits since the X-ray diffraction patterns for the cycled and uncycled electrodes are identical [208].

On deposition of a thin palladium film on gold, Rand and Woods [201] observed an oxide reduction peak corresponding to the gold substrate and a second peak at a more cathodic potential corresponding to an alloy composition of the surface. The oxide reduction peak for the pure palladium oxide was not observed. This result thus indicates the formation of an alloy by deposition of palladium on gold; such phenomenon was also observed for the deposition of rhodium on platinum [201], as mentioned above.

From the oxide reduction peak potential, Rand and Woods [201] estimate the alloy composition as 53% atomic palladium for a 30-sec plating time.

Similar results were obtained by Cadle [207], who observed, for very thin palladium deposits on gold, an oxide reduction peak at a more anodic potential than that for the pure palladium metal. As observed by other authors [201,208], the potential cycling causes a preferential dissolution of palladium and after several cycles the plated electrode behaves like a pure gold electrode.

E. Other Noble Metal Alloys

In this section we briefly mention other noble metal alloys for which the experimental data on the anodic oxide formation and reduction are rather scarce.

The cyclicvoltammogram for a 75% palladium, palladium-rhodium smooth alloy is shown in Figure 39 [203]. For this alloy also, the oxide reduction peak occurs at a potential value intermediate between those for the pure metal constituents, thus suggesting an homogeneous alloy formation. Upon cycling, this potential peak does not shift indicating an homogeneous alloy dissolution [203]. The increase in peak current upon cycling is related to a roughening of the electrode. The analysis of the solution after cycling indicated an equivalent dissolution of both palladium and rhodium confirming the lack of alteration in the surface composition by this electrochemical treatment [203]. In contrast to the case of the smooth alloy, deposited palladium film on rhodium does not indicate the alloy formation as in the case of palladium on gold and rhodium on platinum. The oxide reduction peaks are characteristic of the pure metals [195].

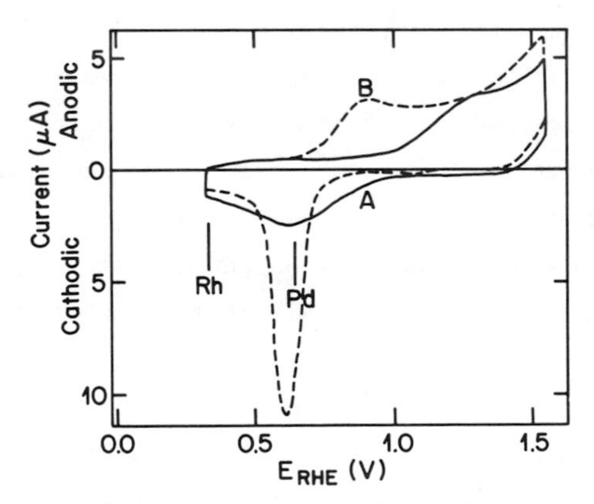

FIG. 39. Cyclicvoltammograms of a 75% palladium, palladium-rhodium alloy; 1st cycle (A) and 1,000th cycle (B); at 25°C, v = 40 mV sec^{-1} in 1 M H_2SO_4. Also indicated are the peak positions for the oxide reduction of the pure metals under identical experimental conditions [203].

The foregoing comments also apply to other deposited noble metals, namely, palladium on platinum and rhodium on gold where the cyclicvoltammograms reflect the additive character of the pure metal without any interaction between the deposited metals.

Electrodeposited iridium on platinum was studied by Aliua and Podlovchenko [209]; for this system iridium dissolves upon potential cycling. No systematic study of oxide reduction as a function of the electrodeposited iridium thickness was performed. The anodic oxidation of such an electrode indicates the additive effect of platinum and iridium [209].

Appleby [210] studied the oxygen reduction on platinum-ruthenium alloys and found no improvements in the oxygen reduction rates upon alloying platinum with ruthenium. An increase in activity in contaminated solutions was found, however, for a low-ruthenium-content alloy [210]. Such an increase in the rate of oxygen reduction was also reported by Hoar and Brooman [211,212], especially at low ruthenium contents (1.9 to 9.3%).

RECENT DEVELOPMENTS

Very recent publications [213-243] indicate a vigorous activity in the field of oxide film formation on platinum [213-228], gold [229-235], palladium [236,237], iridium [238-240], and ruthenium [241-243]. These studies indicate that some controversial aspects of the subject are far from being settled. In Gilroy's work [224], in which a very comprehensive set of data obtained by the potentiostatic potential step procedure were reported on theoxide film formation and reduction, the limiting coverage for platinum oxide growth was not observed. Gilroy [224] concludes that such a limiting coverage [218,219] is an experimental artifact. This is certainly debatable since this behavior has been reported in numerous publications and again very recently under varying experimental conditions [218,219]. The optical and spectroscopic investigation on oxide film formation on platinum and gold have also been reported [214,215,221,222,229].

For complete references, see Addendum to Chapter 1 on page 167.

ACKNOWLEDGMENTS

The authors would like to thank Dr. L. Boulet, the founding Director of Hydro-Quebec Institute of Research (IREQ) for providing facilities and atmosphere conducive to the pursuit of scientific work. Grateful acknowledgment is made also to Dr. G.-G. Cloutier, Assistant Director of IREQ, and Dr. R. Bartnikas, Scientific Director, Materials Science Department of IREQ, for facilities and encouragement.

REFERENCES

1. J. P. Hoare, *The Electrochemistry of Oxygen*, Interscience, New York, 1968; idem, in *Advances in Electrochemistry and Electrochemical Engineering*, Vol. 6 (P. Delahay, ed.), Interscience, New York, 1967.

2. S. Gilman, *Electroanalytical Chemistry*, Vol. 2 (A. J. Bard, ed.), Marcel Dekker, New York, 1967.

3. S. Gilman, *Electrochim. Acta*, *9*, 1025 (1964).

4. F. Will and C. Knorr, *Z. Elektrochem.*, *64*, 258 (1960).

5. M. Becker and M. W. Breiter, *Z. Elektrochem.*, *60*, 1080 (1950).

6. H. A. Laitinen and C. G. Enke, *J. Electrochem. Soc.*, *107*, 773 (1960).

7. W. Visscher and M. A. V. Devanathan, *J. Electroanal. Chem.*, *8*, 127 (1964).

8. W. Böld and M. W. Breiter, *Electrochim. Acta*, *5*, 145 (1961).

9. H. Dietz and H. Göhr, *Electrochim. Acta*, *8*, 343 (1963).

10. K. Vetter and D. Berndt, *Z. Elektrochem.*, *62*, 378 (1958).

11. S. Schuldiner and T. B. Warner, *J. Electrochem. Soc.*, *112*, 212 (1965).

12. S. W. Feldberg, C. G. Enke, and C. E. Bricker, *J. Electrochem. Soc.*, *110*, 826 (1963).

13. K. J. Vetter and J. W. Schultze, *J. Electroanal. Chem.*, *34*, 131 (1972).

14. K. J. Vetter and J. W. Schultze, *J. Electroanal. Chem.*, *34*, 141 (1972).

15. J. W. Schultze, *Z. Phys. Chem.*, NF, *73*, 29 (1970).

16. D. D. Eley and P. R. Wilkinson, *Proc. Roy. Soc.* (London), *A254*, 327 (1961).

17. F. P. Fehlner and N. F. Mott, *Oxidation of Metals*, *2*, 59 (1970).

18. H. Angerstein-Kozlowska, B. E. Conway, and W. B. A. Sharp, *J. Electroanal. Chem.*, *43*, 9 (1973).

19. B. V. Tilak, B. E. Conway, and H. Angerstein-Kozlowska, *J. Electroanal. Chem.*, *48*, 1 (1973).

20. M. D. Goldstein, T. I. Zalkind, and V. I. Veselovskii, *Elektrokhim.*, *10*, 1533 (1974).

21. T. Biegler, *Aust. J. Chem.*, *26*, 2571 (1973).

22. D. Gilroy and B. E. Conway, *Can. J. Chem.*, *46*, 875 (1968).

23. T. Biegler and R. Woods, *J. Electroanal. Chem.*, *35*, 73 (1969).

24. D. A. J. Rand and R. Woods, *J. Electroanal. Chem.*, *35*, 209 (1972).

25. S. Srinivasan and E. Gileadi, *Electrochim. Acta*, *11*, 321 (1966).

26. J. S. Mayell and S. H. Langer, *J. Electrochem. Soc.*, *111*, 438 (1965).

27. M. Fleischmann, I. R. Mansfield, and W. F. K. Wynne-Jones, *J. Electroanall Chem.*, *10*, 511 (1965).

28. A. K. Vijh and B. E. Conway, *Z. Anal. Chem.*, *224*, 160 (1967).

29. Yu. M. Tyurin and G. F. Volodin, *Elektrokhim.*, *5*, 1203 (1969).

30. Yu. M. Tyurin and G. F. Volodin, *Elektrokhim.*, *6*, 1186 (1970).

31. C. A. Khanova, E. V. Kasatkin, and V. I. Veselovskii, *Elektrokhim.*, *8*, 451 (1972).

32. J. Balej and O. Spalek, *Collection Czechoslov. Chem. Commun.*, *37*, 499 (1972).

33. R. Woods, *J. Electroanal. Chem.*, *21*, 457 (1969).

34. W. Visscher and M. Blijlevens, *J. Electroanal. Chem.*, *47*, 363 (1973).

35. R. Parsons and W. H. M. Visscher, *J. Electroanal. Chem.*, *36*, 329 (1972).

36. J. W. Schultze and K. J. Vetter, *Ber. Bunsenges. Phys. Chem.*, *75*, 470 (1971).

37. V. S. Bagotskii, W. J. Lukjanytschewa, A. J. Osche, and W. J. Tichomirov, *Dokl. Akad. Nauk. SSSR*, *159*, 644 (1964).

38. M. A. Dembrowskii, J. M. Kolotyrkin, A. N. Tschomodanaw, and T. W. Kudrjawina, *Dokl. Akad. Nauk. SSSR*, *171*, 1384 (1966).

39. T. Biegler, *J. Electrochem. Soc.*, *116*, 1130 (1969).

40. S. Bruckenstein, D. C. Johnson, and D. T. Napp, *Electrochim. Acta*, *15*, 1493 (1970).

41. J. O'M. Bockris, M. A. Genshaw, and A. K. N. Reddy, *J. Electroanal. Chem.*, *8*, 406 (1964).

42. J. L. Ord and F. C. Ho, *J. Electrochem. Soc.*, *118*, 46 (1971).

43. D. Inman and M. J. Weaver, *J. Electroanal. Chem.*, *51*, 45 (1974).

44. A. Damjanovic, A. T. Ward, B. Ulrick, and M. O'Jea, *J. Electrochem. Soc.*, *122*, 471 (1975).

45. L. B. Harris and A. Damjanovic, *J. Electrochem. Soc.*, *122*, 593 (1975).

46. P. Stonehart, H. A. Kozlowska, and B. E. Conway, *Proc. Roy. Soc.*, (London), *A310*, 541 (1960).

47. A. J. Calandra, N. R. De Tacconi, and A. J. Arvia, *J. Electroanal. Chem.*, *49*, 145 (1974).

48. N. R. De Tacconi, A. J. Calandra, and A. J. Arvia, *J. Electroanal. Chem.*, *51*, 25 (1974).

49. N. R. De Tacconi, A. J. Calandra, and A. J. Arvia, *J. Electroanal. Chem.*, *57*, 325 (1974).

50. N. R. De Tacconi, A. J. Calandra, and A. J. Arvia, *J. Electroanal. Chem.*, *57*, 267 (1974).

51. T. Biegler, D. A. J. Rand, and R. Woods, *J. Electroanal. Chem.*, *29*, 269 (1971).

52. E. Momot and G. Bronoël, *C.R. Acad. Sc. Paris*, *275*, 721 (1972).

53. E. Momot and G. Bronoël, *J. Chim. Phys.*, *70*, 1651 (1973).

54. D. E. Icenhower, H. B. Urbach, and J. A. Harrison, *J. Electrochem. Soc.*, *117*, 1500 (1970).

55. B. Ershler, *Discussion Faraday Soc.*, *1*, 269 (1947).

56. S. Shibata, *Bull. Chem. Soc. Japan*, *36*, 525 (1963).

57. S. Shibata, *Bull. Chem. Soc. Japan*, *40*, 696 (1967).

58. A. Kozawa, *J. Electroanal. Chem.*, *8*, 20 (1964).

59. S. D. James, *J. Electrochem. Soc.*, *116*, 1681 (1969).

60. S. Shibata and M. P. Sumino, *Electrochim. Acta*, *16*, 1089 (1971).

61. W. Visscher and M. Blijlevens, *Electrochim. Acta*, *19*, 387 (1974).

62. S. Shibata, *Electrochim. Acta*, *17*, 395 (1972).

63. S. Shibata and M. P. Sumino, *Electrochim. Acta*, *20*, 871 (1975).

64. B. Warner and S. Schuldiner, *J. Electrochem. Soc.*, *112*, 853 (1965).

65. R. Thacker and J. P. Hoare, *J. Electroanal. Chem.*, *30*, 1 (1971).

66. J. P. Hoare, R. Thacker, and C. R. Wiese, *J. Electroanal. Chem.*, *30*, 15 (1971).

67. J. P. Hoare, *J. Electrochem. Soc.*, *121*, 872 (1974).

68. J. P. Hoare, *Electrochim. Acta*, *20*, 267 (1975).

69. A. J. Appleby, *J. Electroanal. Chem.*, *35*, 193 (1972).

70. D. A. J. Rand and R. Woods, *J. Electroanal. Chem.*, *47*, 353 (1973).

71. L. Formaro and S. Trasatti, *Electrochim. Acta*, *12*, 1457 (1967).

72. J. P. Hoare, *Nature*, *204*, 71 (1964).

73. J. P. Carr and N. A. Hampson, *Electrochim. Acta*, *17*, 2117 (1972).

74. K. Ohashi, K. Sasaki, and S. Nagaura, *Bull. Chem. Soc. Japan*, *34*, 2066 (1966).

75. A. N. Frumkin, E. I. Kruscheva, R. Turasevich, and N. A. Shumilova, *Elektrokhim.*, *1*, 17 (1965).

76. C. C. Liang and A. L. Juliard, *J. Electroanal. Chem.*, *9*, 390 (1965).

77. P. G. Peters and R. A. Mitchell, *J. Electroanal. Chem.*, *10*, 306 (1965).

78. G. Bianchi and T. Mussini, *Electrochim. Acta*, *10*, 445 (1965).

79. M. Ioi and K. Sasaki, *Bull. Chem. Soc. Japan*, *41*, 1028 (1968).

80. A. J. Appleby, *J. Electrochem. Soc.*, *120*, 1205 (1973).

81. K. Sasaki and K. Ohashi, *Electrochim. Acta*, *12*, 366 (1967).

82. J. T. Lundquist and P. Stonehart, *Electrochim. Acta*, *18*, 349 (1973).

83. J. Bett, J. Lundquist, E. Washington, and P. Stonehart, *Electrochim. Acta*, *18*, 343 (1973).

84. K. F. Blurton, P. Greenberg, H. G. Oswin, and D. R. Rutt, *J. Electrochem. Soc.*, *119*, 559 (1972).

85. B. E. Conway, in *Techniques of Electrochemistry*, Vol. 1 (E. Yeager and A. J. Salkind, eds.), Wiley-Interscience, New York, 1972.

86. A. K. N. Reddy, M. A. Genshaw, and J. O'M. Bockris, *J. Electroanal. Chem.*, *8*, 406 (1964).

87. A. K. N. Reddy, M. A. Genshaw, and J. O'M. Bockris, *J. Chem. Phys.*, *48*, 671 (1968).

88. R. Greef, *J. Chem. Phys.*, *51*, 3148 (1969).

89. M. A. Genshaw and J. O'M. Bockris, *J. Chem. Phys.*, *51*, 3149 (1969).

90. R. Greef, *Rev. Scient. Instr.*, *41*, 532 (1970).

91. Y. Vinnikov, V. A. Shepelin, and V. I. Veselovskii, *Elektrokhim.*, *9*, 552 (1973).

92. M. A. Barrett and R. Parsons, *Symp. Faraday Soc.*, *4*, 72 (1970).

93. B. E. Conway, *Symp. Faraday Soc.*, *4*, 95 (1970).

94. F. Chao, M. Costa, and A. Tadjeddine, *Bull. Soc. Chim. France*, 2465 (1971).

95. S. H. Kim, W. Paik, and J. O'M. Bockris, *Surface Sci.*, *33*, 617 (1972).

96. W. Visscher, *Optik*, *26*, 403 (1967).

97. A. Damjanovic, A. T. Ward, and M. O'Jea, *J. Electrochem. Soc.*, *121*, 1186 (1974).

98. Y-C. Chiu and M. A. Genshaw, *J. Phys. Chem.*, *73*, 3571 (1969).

99. J. Horkans, B. D. Cahan, and E. Yeager, *Int. Symp. on Characterization of Adsorbed Species in Catalytic Reactions*, Extended Abstracts, Univ. Ottawa, 1974, p. 18.

100. J. Horkans, B. D. Cahan, and E. Yeager, *Surface Sci.*, *46*, 1 (1974).

101. R. Parsons and W. H. M. Visscher, *J. Electroanal. Chem.*, *36*, 329 (1972).

102. Y. Y. Vinnikov, V. A. Shepelin, and V. I. Veselovskii, *Elektrokhim.*, *9*, 649 (1973).

103. Y. Y. Vinnikov, V. A. Shepelin, and V. I. Veselovskii, *Elektrokhim.*, *9*, 1557 (1973).

104. B. E. Conway and S. Gottesfeld, *J. Chem. Soc.*, *Faraday Trans.*, *1*, , 1090 (1973).

105. M. A. Barrett and R. Parsons, *J. Electroanal. Chem.*, *42*, App. 1 (1973).

106. J. D. E. McIntyre and P. M. Kolb, *Symp. Faraday Soc.*, *4*, 99 (1970).

107. R. M. Lazorenko-Manevich and T. N. Stoyanovskaya, *Elektrokhim.*, *8*, 982 (1972).

108. A. Bewick and A. M. Tuxford, *Symp. Faraday Soc.*, *4*, 114 (1970).

109. J. D. E. McIntyre and D. E. Aspnes, *Surface Sci.*, *24*, 217 (1971).

110. K. S. Kim, N. Winograd, and R. E. Davis, *J. Amer. Chem. Soc.*, *93*, 6296 (1971).

111. G. C. Allen, P. M. Tucker, A. Capon, and R. Parsons, *J. Electroanal. Chem.*, *50*, 335 (1974).

112. T. Dickinson, A. F. Povery, and P. M. A. Sherwood, *J. Chem. Soc.*, *Faraday Trans. I*, *71*, 298 (1975).

113. C. M. Bancroft, I. Adams, L. L. Coatsworth, C. D. Bennewitz, J. D. Brown, and W. D. Westwood, *Anal. Chem.*, *47*, 586 (1975).

114. W. N. Deglass, T. R. Hughes, and C. S. Fadley, *Cat. Rev.*, *4*, 179 (1970).

115. W. J. Johnson and L. A. Heldt, *J. Electrochem. Soc.*, *121*, 34 (1974).

116. S. Shibata, *Electrochim. Acta*, *17*, 395 (1972).

117. C. C. Schubert, C. L. Page, and B. Ralph, *Electrochim. Acta*, *18*, 33 (1973).

118. M. J. Weaver, *J. Electroanal. Chem.*, *51*, 231 (1974).

119. R. P. Nadebaum and T. Z. Fahidy, *Electrochim. Acta*, *17*, 1659 (1972).

120. A. Hickling, *Trans. Faraday Soc.*, *42*, 518 (1946).

121. F. G. Will and C. A. Knorr, *Z. Elektrochem.*, *64*, 270 (1960).

122. S. E. S. El Wakkad, and A. M. S. El Din, *J. Chem. Soc.*, *3098* (1954).

123. S. B. Brummer and A. C. Makrides, *J. Electrochem. Soc.*, *111*, 1122 (1964).

124. S. B. Brummer, *J. Electrochem. Soc.*, *112*, 633 (1965).

125. H. A. Laitinen and M. S. Chao, *J. Electrochem. Soc.*, *108*, 726 (1961).

126. J. W. Schultze and K. J. Vetter, *Ber. Bunsenges Phys. Chem.*, *75*, 470 (1971).

127. D. Dickertmann, J. W. Schultze, and K. J. Vetter, *J. Electroanal. Chem.*, *55*, 429 (1974).

128. A. Hamelin and M. Sotto, *C.R. Acad. Sc. Paris*, *271*, 609 (1970).

129. M. Sotto, *C.R. Acad. Sc. Paris*, *274*, 1776 (1972).

130. A. Capon and R. Parsons, *J. Electroanal. Chem.*, *39*, 275 (1972).

131. C. M. Ferro, A. J. Calandra, and A. J. Arvia, *J. Electroanal. Chem.*, *50*, 403 (1974).

132. C. M. Ferro, A. J. Calandra, and A. J. Arvia, *J. Electroanal. Chem.*, *55*, 291 (1974).

133. C. M. Ferro, A. J. Calandra, and A. J. Arvia, *J. Electroanal. Chem.*, *59*, 239 (1975).

134. C. M. Ferro, A. J. Calandra, and A. J. Arvia, *J. Electroanal. Chem.*, *65*, 963 (1975).

135. S. H. Cadle and S. Bruckenstein, *Anal. Chem.*, *46*, 16 (1974).

136. M. D. Goldshtein, T. I. Zalkind, and V. I. Veselovskii, *Elektrokhim.*, *8*, 606 (1972).

137. M. D. Goldshtein, T. I. Zalkind, and V. I. Veselovskii, *Elektrokhim.*, *9*, 699 (1973).

138. G. Grüneberg, *Electrochim. Acta*, *10*, 339 (1965).

139. D. A. J. Rand and R. Woods, *J. Electroanal. Chem.*, *31*, 29 (1971).

140. K. Moslavac, B. Lovrecek, and R. Radeka, *Electrochim. Acta*, *17*, 415 (1972).

141. J. P. Hoare, *Electrochim. Acta*, *9*, 1289 (1964).

142. J. P. Hoare, *Electrochim. Acta*, *11*, 311 (1966).

143. R. A. Bonewite and G. M. Schmid, *J. Electrochem. Soc.*, *117*, 1367 (1970).

144. W. E. Reid and J. Kruger, *Nature*, *203*, 402 (1964).

145. R. S. Sirohi and M. A. Genshaw, *J. Electrochem. Soc.*, *116*, 910 (1969).

146. Y. Y. Vinnikov, V. A. Shepelin, and V. I. Veselovskii, *Elektrokhim.*, *8*, 1229 (1972).

147. Y. Y. Vinnikov, V. A. Shepelin, and V. I. Veselovskii, *Elektrokhim.*, *8*, 1384 (1972).

148. J. L. Ord and D. J. De Smet, *J. Electrochem. Soc.*, *118*, 206 (1971).

149. B. E. Conway, L. H. Laliberté, and S. Gottesfeld, in *Proc. Symp. Oxide-Electrolyte Interface*, Electrochemical Society, Princeton, New Jersey, 1973.

150. T. Takamura, K. Takamura, W. Nippe, and E. Yeager, *J. Electrochem. Soc.*, *117*, 626 (1970).

151. R. M. Lazorenko-Manevich and T. N. Stoyanovskaya, *Elektrokhim.*, *8*, 1113 (1972).

152. K. S. Kim, C. D. Sell, and N. Winograd, in *Electrocatalysis* (M. W. Breiter, ed.), Electrochemical Society, Princeton, New Jersey, 1974.

153. J.-M. Cesbron, R. Courtel, J.-E. Dubois, M. Herlem, and P.-C. Lacaze, *C.R. Acad. Sc. Paris*, *266*, 1667 (1968).

154. J.-M. Cesbron, R. Courtel, J.-E. Dubois, and P.-C. Lacaze, *C.R. Acad. Sc. Paris*, *268*, 1985 (1969).

155. J. E. Dubois, P. C. Lacaze, R. Courtel, C. C. Herrmann, and D. Maugis, *J. Electrochem. Soc.*, *122*, 1454 (1975).

156. S. E. S. El Wakkad and A. M. S. El Din, *J. Chem. Soc.*, *3094* (1954).

157. A. Hickling and G. Vrjosek, *Trans. Faraday Soc.*, *57*, 123 (1961).

158. T. R. Blackburn and J. J. Lingane, *J. Electroanal. Chem.*, *5*, 216 (1963).

159. F. G. Will and C. A. Knorr, *Z. Elektrochem.*, *64*, 270 (1960).

160. J. P. Hoare, *J. Electrochem. Soc.*, *111*, 610 (1964).

161. S. Schuldiner and R. M. Roe, *J. Electrochem. Soc.*, *111*, 369 (1964).

162. S. H. Cadle, *J. Electrochem. Soc.*, *121*, 645 (1974).

163. J. F. Llopis, J. M. Gamboa, and L. Victori, *Electrochim. Acta*, *17*, 2225 (1972).

164. K. S. Kim, A. F. Gossmann, and N. Winograd, *Anal. Chem.*, *46*, 197 (1974).

165. M. R. Tarasevich, V. S. Vilinskaya, and R.Kh. Burshtein, *Elektrokhim.*, *7*, 1200 (1971).

166. S. Shibata, *Bull. Chem. Soc. Japan*, *38*, 1330 (1965).

167. W. Böld and M. W. Breiter, Electrochim. Acta, 5, 169 (1961).

168. A. K. Vijh, *Electrochemistry of Metals and Semiconductors*, Marcel Dekker, New York, 1973, p. 129.

169. J. P. Hoare, *J. Electrochem. Soc.*, *111*, 232 (1964).

170. R. Kh. Burshtein, M. R. Tarasevich, and K. A. Radyushkina, *Electrokhim.*, *6*, 1611 (1970).

171. E. I. Khrushcheva, N. A. Shumilova, and M. R. Tarasevich, *Elektrokhim.*, *2*, 277 (1966).

172. J. Llopis and M. Vazquez, *Electrochim. Acta*, *9*, 1655 (1964).

173. W. Böld and M. W. Breiter, *Electrochim. Acta*, *5* 169 (1961).

174. M. W. Breiter, *Z. Phys. Chem. NF*, *52*, 73 (1967).

175. B. D. Kurnikov, A. I. Zhurin, V. V. Chernyi, Yu. B. Vasil'ev, and V. S. Bagotskii, *Elektrokhim.*, *9*, 833 (1973).

176. D. A. J. Rand and R. Woods, *J. Electroanal. Chem.*, *55*, 375 (1974).

177. J. M. Otten and W. Visscher, *J. Electroanal. Chem.*, *55*, 1 (1974).

178. J. M. Otten and W. Visscher, *J. Electroanal. Chem.*, *55*, 13 (1974).

179. A. Damjanovic, A. Dey, and J. O'M. Bockris, *J. Electrochem. Soc.*, *113*, 739 (1966).

180. B. I. Podlovchenko and N. A. Epshtein, *Elektrokhim.*, *9*, 1194 (1973).

181. B. D. Kurnikov and Yu. B. Vasil'ev, *Elektrokhim.*, *9*, 1203 (1973).

182. B. D. Kurnikov and Yu. B. Vasil'ev, *Elektrokhim.*, *9*, 1739 (1973).

183. A. V. Boiko and L. I. Kadaner, *Elektrokhim.*, *9*, 1357 (1973).

184. B. D. Kurnikov and Yu. B. Vasil'ev, *Elektrokhim.*, *10*, 77 (1974).

185. J. Llopis, I. M. Tordesillas, and J. M. Alfayate, *Electrochim. Acta*, *11*, 623 (1966).

186. J. Llopis, J. M. Gamboa, and J. M. Alfayate, *Electrochim. Acta*, *12*, 57 (1967).

187. V. V. Gorodetskii, M. M. Pecherskii, Ya. B. Skuratnik, M. A. Dombrovskii, and V. V. Losev, *Elektrokhim.*, *9*, 894 (1973).

188. S. Trasatti and G. Buzzanca, *J. Electroanal. Chem.*, *29*, App. 1 (1971).

189. D. Galizzioli, F. Tantardini, and S. Trasatti, *J. Appl. Electrochem.*, *4*, 57 (1974).

190. S. Hadzi-Jordanov, H. Angerstein-Kozlowska, and B. E. Conway, *J. Electroanal. Chem.*, *60*, 359 (1975).

191. A. T. Kuhn and P. M. Wright, *J. Electroanal. Chem.*, *41*, 329 (1973).

192. W. O'Grady, C. Iwakura, J. Huang, and E. Yeager, in *Electrocatalysis* (M. W. Breiter, ed.), Electrochemical Society, Princeton, New Jersey, 1974.

193. L. D. Burke and T. O. O'Meara, *J. Chem. Soc.*, *Faraday Trans. I*, *68*, 839 (1972).

194. G. P. Khomchenko, N. G. Ul'ko, and G. D. Vovchenko, *Elektrokhim.*, *1*, 659 (1965).

195. A. J. Appleby, *J. Electrochem. Soc.*, *117*, 1157 (1970).

196. J. Llopis and M. Vazquez, *Anales Real Soc. Espan. Fiz. Quim. (Madrid)*, *1363*, 273 (1967).

197. R. Woods, *Electrochim. Acta*, *16*, 655 (1971).

198. M. W. Breiter, *J. Phys. Chem.*, *69*, 901 (1965).

199. K. A. Radyushkina, R. Kh. Burshtein, M. R. Tarasevich, and V. V. Kuprina, *Electrokhim.*, *6*, 234 (1970).

200. A. A. Michri, A. G. Pshenichnikov, R. Kh. Burshtein, and V. B. Bernard, *Electrokhim.*, *6*, 719 (1970).

201. D. A. J. Rand and R. Woods, *J. Electroanal. Chem.*, *44*, 83 (1973).

202. R. Woods, *Electrochim. Acta*, *14*, 533 (1969).

203. D. A. J. Rand and R. Woods, *J. Electroanal. Chem.*, *36*, 57 (1972).

204. J. S. Mayell and W. A. Barber, *J. Electrochem. Soc.*, *116*, 1333 (1964).

205. A. A. Michri, A. G. Pshenichnikov, R. Kh. Burshtein, and V. B. Bernard, *Elektrokhim.*, *5*, 603 (1969).

206. K. A. Radyushkina, R. Kh. Burshtein, M. R. Tarasevich, V. V. Kuprina, and L. A. Cheriqaev, *Elektrokhim.*, *5*, 1379 (1969).

207. S. H. Cadle, *Anal. Chem.*, *46*, 587 (1974).

208. R. Woods, *Electrochim. Acta*, *14*, 632 (1969).

209. L. Aliua and B. I. Podlovchenko, *Elektrokhim.*, *9*, 1215 (1973).

210. A. J. Appleby, *J. Electroanal. Chem.*, *27*, 347 (1970).

211. T. P. Hoar and E. W. Brooman, *Electrochim. Acta*, *11*, 545 (1969).

212. E. W. Brooman and T. P. Hoar, *Platinum Metals Rev.*, *9*, 122 (1966).

Chapter 2

GAS DISCHARGE ANODIZATION

John F. O'Hanlon

IBM Corporation
Thomas J. Watson Research Center
Yorktown Heights, New York

I. INTRODUCTION

Gas discharge anodization, a technique for the formation of oxides on metals, is most easily described as oxide formation in an electrolytic cell in which the liquid electrolyte has been replaced by some form of an oxygen discharge. Work in this system was stimulated by a desire to understand anodic film growth in an environment not complicated by the effects of electrolyte incorporation or film dissolution, and continued because of the promise of a low-temperature process which might be clean and compatible with other vacuum processing steps in semiconductor device technology.

Early work on the oxidation of selenium in an oxygen discharge by Olsen and Meloche [1] and the oxidation of silver in an air corona by Gunterschultze and Betz [2] set the stage for the pioneering work in plasma anodization by Dankov and Ignatov [3], Ignatov [4], and Tiapkina and Dankov [5] in 1946. In those studies films were made on the anode of a direct current (DC) glow discharge. Following this work many techniques including DC biasing of the sample [6], radio frequency (RF) [7], microwave [8], and DC arc [9] discharges evolved. Applications as diverse as capacitors, Josephson junctions, xerography, and planar transistors have led to an interest in understanding the process and its limitations.

This chapter presents a review of the major techniques for producing gaseous electrolytes and anodic oxides, comments on the growth mechanisms and finally, surveys the properties of a large number of oxides fabricated by these techniques.

II. OXYGEN DISCHARGE PROPERTIES

The understanding of the oxide growth parameters that are presented in the following sections can be made clear by first examining some of the physical and geometrical properties of the oxygen discharges that have been used to carry out gas-phase anodization. Many experimental effects observed during gas anodization are a natural result of either the type of discharge employed, or one of the many parametric effects such as geometry, temperature, or surface catalysis. In this section the characteristics of several oxygen plasmas are reviewed.

A. DC Discharges

1. *The DC Glow Discharge*

The general characteristics of the various regions of a glow discharge have been described in detail elsewhere [10,11] and will only be briefly reviewed here. The cathode fall is a region of the glow in which large fields accelerate electrons into the negative glow (NG) where they make ionizing collisions with neutral molecules. As the electric field in the NG is small, these ions move by diffusion to the edge of the cathode fall where they are accelerated toward the cathode. The impact of these ions on the cathode causes ejection of secondary electrons which complete the sustaining cycle. Only these two regions are necessary for the maintenance of the discharge. As the discharge electrodes are moved farther apart, the mode assumed by the glow will depend upon several parameters; in particular, pressure, current density, spatial confinement of the glow, and electrode area.

Separating the electrodes while continuing to confine the discharge to a cylindrical geometry reveals first the Faraday dark space (FDS) followed by the positive column (PC). The FDS is dark only in a relative sense; here the primary electrons with their energies diminished by collisions are only slightly capable of ionizing neutrals. The PC is a zone of moderate field strength which accelerates electrons to a velocity great enough to produce ionizing collisions. Electron transport in both the FDS and PC is diffusion controlled; these regions serve as electrical connections between the NG and the anode.

The NG of an oxygen discharge operating at a pressure of 0.04-0.08 Torr is characterized by an electron density of $\sim 10^8$-10^9 per cm^3 with an electron temperature of 1.5-0.5 eV. With the aid of a negative-ion mass spectrometer, Thompson [12] has found the concentration of negative ions in the NG to be about 1% of the positive ion concentration, while Lunt and Gregg [13] observed no negative atomic or molecular ions in the NG to a sensitivity limit of $\sim$1% $[O_2^+]$. Drost et al. [14] were able to extract a small (10^{-10} A/cm^2) negative ion current from the NG; however, potential effects may have made the electrical aperture smaller than the physical aperture.

The character of the PC is rather different from that of the NG. Here the electron density is of order $10^7/cm^3$, with an average electron temperature of 3-4 eV [15] (Fig. 1). Thompson [12] found the negative ion concentration to be 20 times greater than the electron concentration and the negative ion energy to be $\sim$0.15 eV. In the PC, negative ions are produced at a rate of $\sim 10^{12}/cm^3$ per sec by the Lozier process [16]

$$O_2 + e \rightarrow O^- + O^+ + e \qquad (1)$$

for 5- to 10-eV electrons. Since the wall potential traps the negative atomic ions in the gas phase, their concentration builds up until they are destroyed at a sufficient rate by a gas-phase process. The major destruction mechanism for O^- is believed to be collisions with atoms [15].

$$O^- + O \rightarrow O_2 + e \qquad (2)$$

Lunt and Gregg [13] found no negative molecular ions in the PC.

The PC of a confined discharge was classified into four modes by Rundle et al. [17]. They distinguished these modes by the type or absence of high frequency oscillations in the potential at the

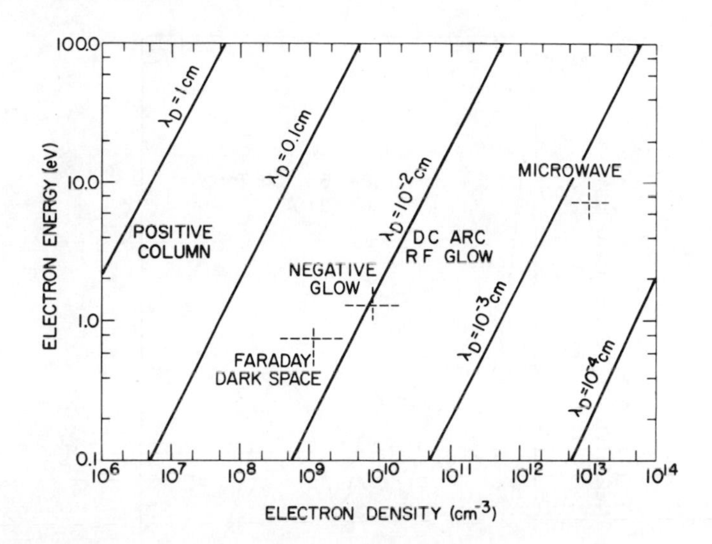

FIG. 1. Electron energy-density regions used in anodization. The Debye length (λ_D) shown is for the case $T_e >> T_i$, $Z_i = 1$.

anode. In the pressure region from 1.5 to 0.5 Torr, the lowest considered in that study, only one mode was observed, which was characterized by a moderate electric field of 60-20 V/cm, depending upon the pressure, with no oscillations in anode potential. Thompson [12] found that for lower pressures, noise-free operation of his hollow cathod discharge was obtained at only one combination of current and pressure (4 mA, 0.04 Torr). One often subdivides this classification of the PC into two groups; uniform, and visibly striated. This division may be arbitrary, as Drost et al. [14] have shown that the negative ion current extracted laterally from the PC varied periodically along the length of the PC even when the striations were not visible to the eye. Figure 2 illustrates the spatial distribution of the negative ion current extracted by Drost et al. for the case where no striations were visible.

In the presence of striations a similar periodic variation of the negative ion current was found. Negative ion maxima were found at points of maximum brightness, while positive ion maxima were

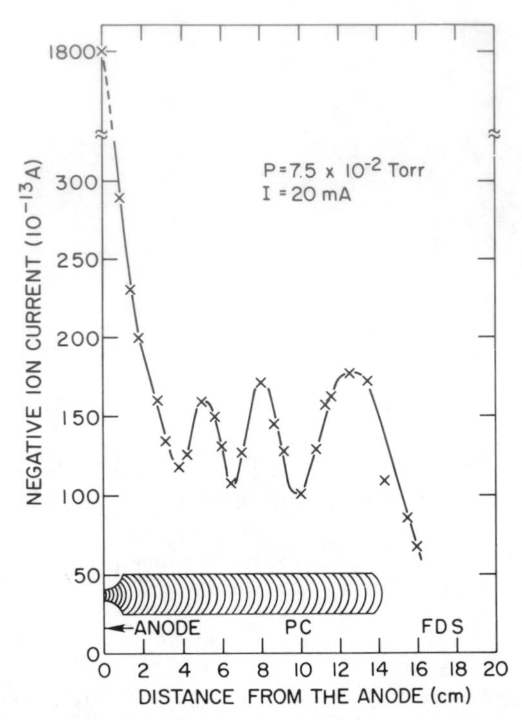

FIG. 2. Intensity curve of extracted negative ion current along a column not visibly striated. After Drost et al. [14].

found to occur at points of minimum brightness and negative ion concentration. The brightest region of all was located directly in front of the anode and corresponded to the largest concentration of negative ions in the discharge. For this type of discharge Rundle et al. [17] found a few percent atomic oxygen distributed uniformly throughout the PC. The concentration of oxygen atoms was found to increase with the discharge current. No ozone was observed. Herron et al. [18] observed 4-12% atomic oxygen in a 60-Hz alternating current (AC) glow which is really an alternating DC glow.

The anode fall is a region of space charge that exists directly in front of the anode. If the area of the anode is large, that is, comparable to the cross-sectional area of the discharge tube, a retarding potential and positive space charge layer develop in order that the random electron flow to the anode be limited to the discharge current. If the area of the anode is made very small, an attractive potential and electronic space charge exist to increase the effective area of the anode so that the random current flow to the anode equals the current needed to maintain the discharge. The magnitude of the attractive potential can be as high as the ionization potential of the gas; in that case a visible glow, the anode glow, appears in front of the anode.

If the cross-sectional area of the discharge is not confined, but is allowed to expand and fill the volume of a large vessel, the PC will disappear and the current will be carried to the anode by diffusion of electrons through the NG. No anode fall develops with large-area cathodes since the diffusion current through the NG is large enough to sustain the discharge. An anode fall will develop only if the area of the anode is made sufficiently small.

2. *The Low Pressure DC Oxygen Arc*

The cathode temperature of the low pressure DC arc is high enough to prevent the cathode from being stable in oxygen for long periods of time. Ligenza and Povilonis [19] developed a tantalum foil cathode surrounded by argon, which was a stable cathode for an oxygen arc column [9]. The arc column is like the PC of the glow discharge in some respects. The optical spectra have many lines in common. The electron temperature is $\sim$3 eV, and neutral atoms are present along with positive molecular ions, but in greater number. The electron density in the arc column is also considerably higher than in the PC, typically 10^{11}-$10^{12}/cm^3$. Due to rapid diffusion of ions and electrons to the wall at low pressures, the arc column expands to fill the discharge chamber.

B. AC Discharges

The development of an AC discharge was characterized by four parameters by Francis [20]: the electron-neutral collision frequency, ν_e (Fig. 3), the electron mean free path, λ_e (Fig. 4), the frequency of the applied electric field, ω, and the size of the discharge vessel, d. The data given in Figures 3 and 4 are based on the elastic scattering cross-section for electrons in oxygen given by McDaniel [21]. For conditions encountered in anodization, three combinations of the above parameters are relevant:

(i) $\lambda_e < d$; $\omega < \nu_e$. In this region the pressure is high enough so that electrons make many collisions before contacting a wall, and therefore drift as a cloud in phase with the applied field. For very low frequencies the cloud of electrons may be driven to the wall each half period; a secondary process at the wall is required to maintain this "alternating DC" discharge.

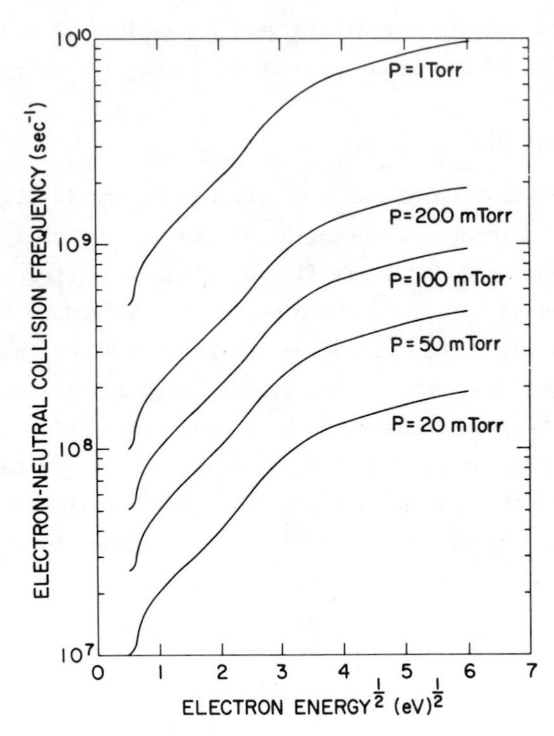

FIG. 3. The electron-neutral oxygen collision frequency (ν_e) for the electron energy-pressure range used in anodic oxidation (temperature = 50°C).

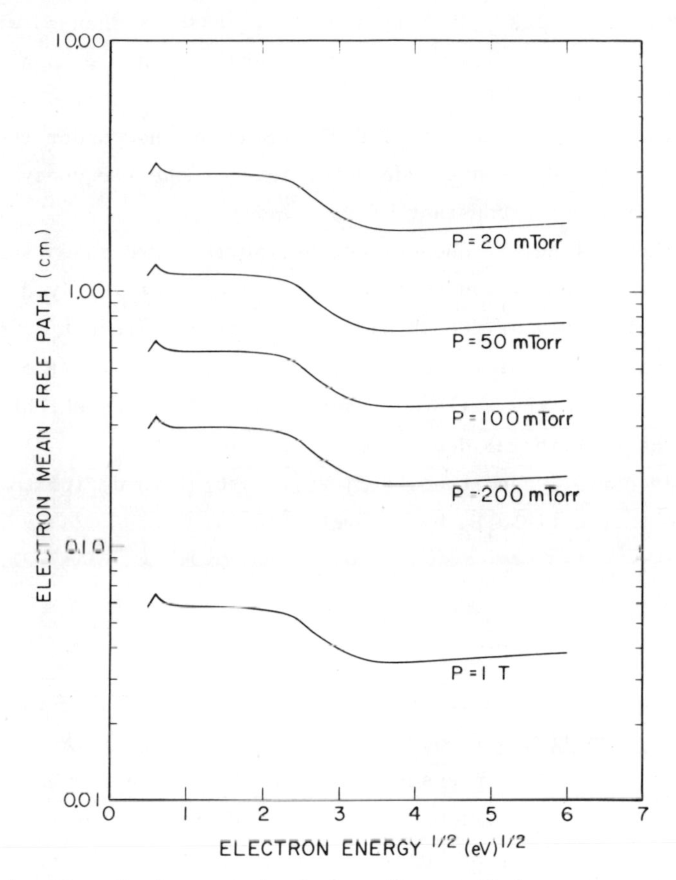

FIG. 4. The electron-neutral mean free path in oxygen at 50°C.

(ii) $\lambda_e < d$; $\omega > \nu_e$. Here the frequency of oscillation is great enough so that the average electron makes many oscillations between collisions. In this regime the electrons slowly gain the energy required to make ionizing collisions and sustain the discharge. The cloud of electrons and ions moves by random diffusion, and in oxygen resembles the PC of a DC glow discharge.

An electron freely oscillating in an electric field cannot gain any energy from the field. In order for electrons to gain any energy from the field they must suffer elastic collisions with neutral gas molecules. In this way their phase is changed with respect to the driving field and they are able to gain enough energy to make ionizing collisions.

(iii) $\lambda_e \sim d$; $\omega >> \nu_e$. The electrons are now under the influence of a standing-wave mode determined by the frequency, geometry, and dielectric constant of the cavity.

In these higher frequency regimes, surface processes such as secondary emission by ion bombardment (γ_i) are not involved. The surfaces act mainly as recombination centers and are not necessary for the maintenance of the discharge. However, the drift of electrons and ions to the walls does set up static fields which largely control the equilibrium density of ionization.

Under certain conditions [22] it is useful to define an effective DC electric field E_e by the relation

$$E_e = \frac{E_a}{(1 + \omega^2/\nu_e^2)^{1/2}} \tag{3}$$

For the usual RF glow discharge (electron energy $\sim$3 eV), the electron-neutral collision frequency is $\sim 7 \times 10^{-8}$ N, where N is the gas density in cm^{-3}. At a pressure of 5×10^{-2} Torr, a temperature of 50°C, and a frequency (ω) of 100 MHz, the magnitudes of ω and ν_e are equal; below about 30 MHz, the rate at which energy is absorbed from the field is independent of frequency.

The starting criterion for an AC discharge is that each electron must create one new electron to replace itself before being lost from the discharge. In a gas containing negative ions this reads,

$$\nu_a n - D\nabla^2 n = \nu_i n \tag{4}$$

The generation of ions, $\nu_i n$, is balanced by losses through electron attachment to atoms or molecules, $\nu_a n$ plus those lost to the walls by diffusion, $D\nabla^2 n$.

The AC oxygen plasmas used in anodization studies have not been thoroughly characterized. Inductively-coupled discharges excited by frequencies as high as 6.7 MHz have been used by Mikhalkin and Odynets [23,24], and Makara et al. [25], while Pulfrey and Reche [26] operated at 0.5-1.0 MHz. Microwave generators operating at 2.5 GHz have been used by Ligenza [8], Kraitchman [27], Skelt and Howels [28], and Weinreich [29] to produce oxygen discharges in the pressure range of 0.1 to 1.5 Torr for anodization. Ligenza [8] estimated the electron temperature to be 6.8 eV and the electron density to be $1.2 \times 10^{13}/cm^3$ in his apparatus.

The concentration of atomic oxygen in the discharge is variable, and due, in part, to catalytic effects; amounts as large as 10-12% have been observed [30]. Elias et al. [30] measured the recombination rate of atoms in a 2.5-GHz discharge, and found that platinum was not a good catalyst for the recombination of oxygen atoms, while the oxides of silver, cobalt, nickel, manganese, and copper were. Large catalytic effects in the production of oxygen atoms have been attributed to hydrogen. Herron and Schiff [18] observed that the concentration of oxygen atoms doubled with the addition of 0.6% water. Kaufman [31] discusses very pure oxygen glows which gave only 0.3% oxygen atoms; the addition to these pure oxygen glows of small amounts of hydrogen (0.01-0.05%) produced 160-200 oxygen atoms per added hydrogen molecule. Similar but smaller effects were seen with nitrogen, nitrous oxide, and nitric oxide. These observations imply that the large recombination term required to describe pure gas discharges is due to surface recombination, and that catalytic dissociation appears to involve a modification of the surface properties [31].

C. Probes

Langmuir probes, when properly used, are well suited for making approximate measurements of the electron energy, density, space potential, and floating potential. The negative potential measured on a small isolated probe drawing no current from the plasma is referred to as the floating potential; it results from the ambipolar diffusion of electrons and ions to the wall. Biasing the probe more negatively than that value results in the formation of a positive ion sheath, while biasing the probe more positively than the floating potential results first in an increase of electron current followed by saturation. The potential of the upper knee of the curve is the space potential, or potential of the plasma with respect to the counter electrode (anode). Above the space potential, an electron sheath forms around the probe.

Waymouth [32] has reviewed the use of probes in plasmas and classified their operation in terms of four variables: the Debye length λ_D, the sheath length d_s, the probe radius r_p, and the electron-neutral mean free path λ_e. For the conditions $\lambda_e > 10r_p$, and $r_p > 100\lambda_D$, the probe, regardless of shape, may be treated as a plane probe using the theory as given by Loeb [33]. For the pressure and electron density ranges of interest in anodization, it is difficult to satisfy both criteria without making the probe so large that it disturbs the plasma. If, however, the second condition is relaxed, the theory of the cylindrical or spherical probe as given by von Engel and Steenbeck [34] or Langmuir and Compton [35] is applicable. For oxygen this requires $r_p < 0.006$ cm for P = 1 Torr, and $r_p < 0.1$ cm for $P = 5 \times 10^{-2}$ Torr.

These concepts are valid for regions of the oxygen discharge which are electropositive, e.g., the NG and FDS, but do not hold in the presence of large numbers of negative ions. The theory of operation of the Langmuir probe in electronegative gases has been developed by Boyd and Thompson [36]. A simple test of the presence of large numbers of negative ions is the measurement of the ratio of saturation electron current i_e to saturation positive ion current i_+. For an electropositive oxygen plasma this ratio is

$$\frac{i_e}{i_+} = \left(\frac{e}{2\pi}\right)^{1/2}\left(\frac{m_+}{m_e}\right)^{1/2} = 159 \qquad (5)$$

while in the PC (where $n_-/n_e \sim 20$), the ratio is an order of magnitude smaller [12].

In gas discharge anodization a sample can exhibit either anode-like or probe-like characteristics. If the sample-probe is small and draws little current from the discharge, it behaves like a true probe; however, if the sample-probe is large or collects a current such that $i_{probe}/i_{discharge}$ is not small, it greatly disturbs the plasma and behaves more like an anode. The area of the probe and the charge flow to it may perturb or greatly change the density and types of particles nearby. A large-area sample-probe also acts as a recombination center further affecting the equilibrium density of certain species. The floating potential of the probe will drift continually with time as material sputtered from the cathode builds up on the anode and shifts the potential energy of the entire plasma. The probe may become contaminated with material sputtered from the cathode or walls of the chamber, or from anodic oxidation when $V_{probe} > V_{floating}$. Since gold was observed to react with the discharge under positive bias, both Nilson et al. [37] and the present author have found it expedient to use platinum probes which were heated prior to each measurement. Even then, the interpretation was complicated by a rapidly varying contact potential. In addition to these complications many other effects can obscure the interpretation of probe data; these have been detailed by Loeb [33].

III. TECHNIQUES FOR OXIDE FORMATION

The majority of the published works on gas-phase anodization were carried out in the low pressure DC glow discharge. This discharge was favored by many because of the experimental simplicity and slow growth rates which made the technique particularly suitable for the production of thin oxides. It is not implied here that the glow was identical in all experimental situations; both the mode of operation and the geometry varied considerably. The remaining oxidation studies were performed in high density plasmas such as the low pressure DC arc, induced RF, and microwave discharges. These studies were stimulated by a desire to obtain higher growth rates for applications requiring thicker films. In this section experimental work representative of each of the many techniques for gas phase anodization is discussed.

A. Low Pressure DC Glow Discharge

1. Oxidation on the Anode

Historically, the first technique to be used for the formation of anodic oxides was also the simplest; the formation of an oxide on the active anode used to maintain a low pressure oxygen discharge. With this technique, Dankov and Ignatov [3] were able to anodize aluminum films and form oxides 10^{-6}-10^{-5} cm thick in about 10-20 min. Metal films formed by condensation on mica were transferred to a grid structure so that the anode became the target of an electron diffraction apparatus. Using this apparatus the complete oxidation of the film to γ'-Al_2O_3 was observed. According to the concepts put forward by Dankov [38], there were two reasons why the film grew thicker with this method than with thermal oxidation at the same temperature: the presence of active particles (e.g., negative ions, atoms, or ozone) possessing a greater electrochemical potential than molecular oxygen, and the presence of an electric field.

Films grown by this technique tend to obey a logarithmic growth law [39,40]. This may be quite fortuitous as is indicated by the cathodic sputter deposition seen in some work [41] and by the lack of constancy of the voltage during anodization.

2. Oxidation on the Cathode

The electric field at the surface of the cathode is in precisely the opposite direction to that needed to cause anodic growth, and yet oxides are still observed to grow (Fig. 5) on the physical cathode used to support the discharge [4,39,41]. This growth is postulated to be due to the impact of energetic ions [4], and local heating of the cathode [41]. Ignatov [39] found the surface temperature rise (200°C) to be incapable of explaining the observed oxide thickness on the basis of thermal oxidation, and concluded that the high energy and activity of the positive ions (O_2^+, O^+) were responsible. On the other hand Nazarova [40] found that a germanium anode, which had previously been anodized until a thick

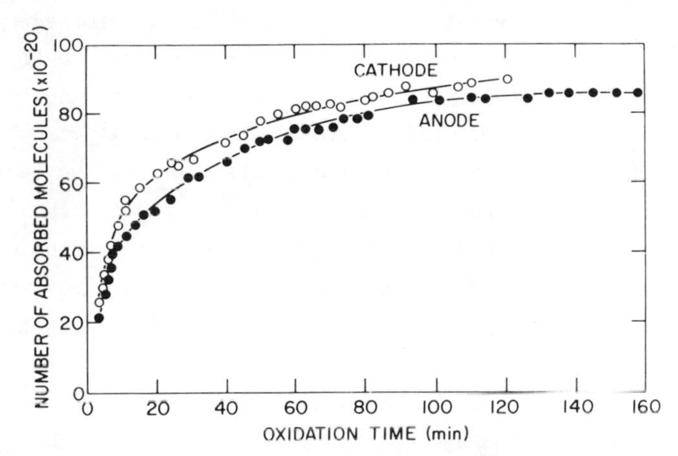

FIG. 5. Oxidation of aluminum on the cathode and anode electrodes in an oxygen glow discharge. After Ignatov [39].

oxide formed, exhibited a continuous weight loss when made the cathode.

The apparent discrepancy in these results can be explained by the work of Greiner [42] on cathodic RF oxidation of lead films and of Scharfe [43] on cathodic DC oxidation of tantalum films. The model evoked by both states that the net rate of oxide growth is the difference between the oxidation rate and the sputter removal rate.

$$\left(\frac{dx}{dt}\right)_{net} = \left(\frac{dx}{dt}\right)_{growth} - \left(\frac{dx}{dt}\right)_{sputtering} \tag{6}$$

In Greiner's work the growth term was enhanced, for example, by increasing the oxygen content of the argon-oxygen discharge, or by raising the substrate temperature, while the sputter removal term increased with increasing discharge potential or argon content. Thus a bare metallic cathode grew an oxide because the oxidation rate of a thin oxide far exceeded the sputter removal rate, while an initially thickly oxidized cathode was sputtered away because the growth rate was less than the sputtering rate. The equilibrium thickness at which the net growth was zero, was determined by a balance between the factors discussed above and varied over a large range. The kinetic balance between growth and sputtering in an oxygen atmosphere is not well understood theoretically.

The results of Greiner are also consistent with the earlier work of Scharfe [43] who grew tantalum oxide cathodically. Scharfe observed that a small piece of metal foil suspended in the discharge about 1 cm away from the cathode allowed the oxide to grow thicker on the cathode region eclipsed by it, by reducing the energetic ion bombardment responsible for sputtering. He also concluded that the final thickness of the oxidation was determined by the relative rate of oxidation and sputtering. In a study of the growth of SiO_2 on a silicon cathode in a DC glow discharge, Lertes [44] found that local cathode sputtering caused pitting, making the films made in his apparatus unusable as dielectric layers.

3. The Oxidation of a Floating Electrode

a. Negative Glow. The oxidation of samples immersed in the discharge, but not electrically connected to a discharge electrode, was described by Miles and Smith [6], who successfully fabricated $Al\text{-}Al_2O_3\text{-}Al$ tunnel junctions. In their study, an aluminum ring cathode electrode was used, while the base plate of the work chamber was the anode (Fig. 6). In this configuration the sample was immersed in the NG. Figure 7 shows the results of measurements of aluminum oxide thicknesses made as a function of the time during which the specimen was left immersed in the oxygen glow discharge. The thickness of the oxide was determined by comparison of the experimentally-observed current-voltage characteristics with electron tunneling theory.

This technique was used by many [45-48] in the early days of tunnel junction fabrication as a clean method of forming thin oxides in a manner compatible with the remainder of the processing steps. This process is not usable for the production of, say, superconductive tunnel junctions as the reproducibility is inadequate.

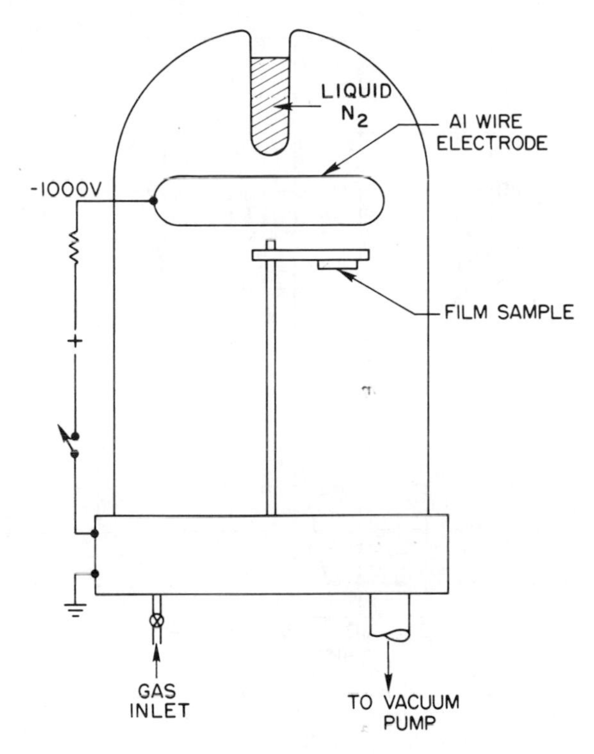

FIG. 6. Schematic diagram of gaseous anodization system used by Miles and Smith. Reprinted from Ref. 6 by courtesy of the Electrochemical Society, Inc.

b. Cathode Fall. The oxidation of electrically floating silicon samples in the cathode fall of a DC glow discharge was studied by Lertes [44]. A silicon wafer was located in the cathode fall, adjacent to but isolated electrically from the cathode by means of a thin quartz wafer. The oxidation was found to be nonuniform (thicker at the edges); the amount of nonuniformity was found to be dependent upon the relative diameters of the silicon wafer, quartz isolation wafer, and the silicon cathode. The radial distribution of oxide thickness with time is shown in Figure 8 for a silicon wafer of 1.5 cm diameter isolated from a silicon cathode of 3 cm diameter. Growth appeared to be logarithmic for the first 3-4 hr and then followed a faster linear growth pattern. The technique is limited by the small useful area of uniform film thickness.

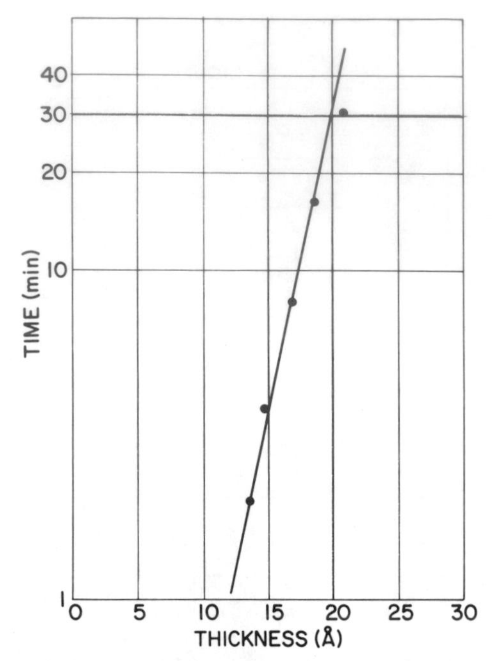

FIG. 7. Thickness versus time for gaseous anodization of an electrically-isolated aluminum film. Reprinted from Ref. 6 by courtesy of the Electrochemical Society, Inc.

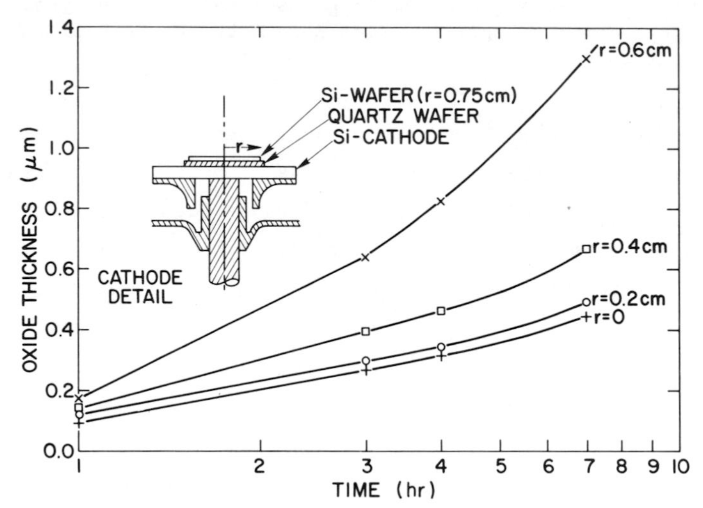

FIG. 8. Radial distribution of oxide thickness with time for silicon oxidized in the cathode fall (after Lertes [44]). The cathode detail is shown in the insert.

4. Oxidation of a Biased Electrode

a. Negative Glow. This technique was first developed by Miles and Smith [6] and applied by them to the oxidation of a wide range of metals. To form an oxide, a film was placed in the NG and a sample voltage more positive than the local plasma potential was applied. The sample voltage was referenced to the anode as it was close in potential to the NG. It was found that the oxide growth rate and sample current became diminishingly small with time until the film assumed an equilibrium thickness. This observation had been carefully reproduced by Nyce [49] who used ellipsometry to monitor the growth of Al_2O_3 on very pure aluminum films deposited on sapphire substrates in ultrahigh vacuum. His results for constant voltage anodization are shown in Figure 9.

If a thicker film is desired, the voltage may be further increased and then held constant while the current is allowed to decay

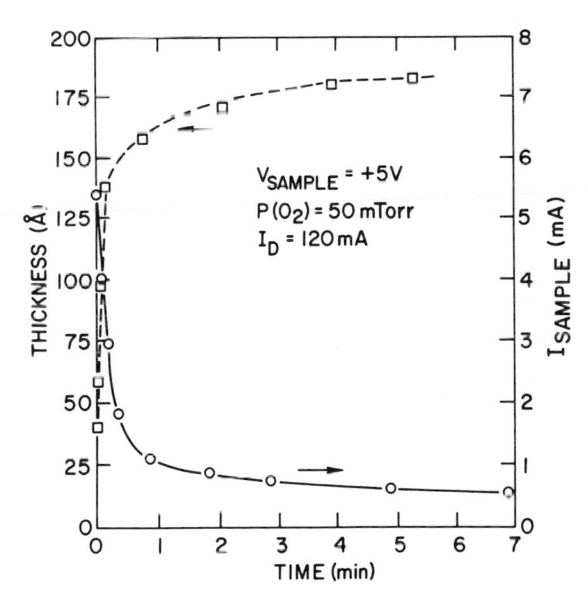

FIG. 9. The growth of Al_2O_3 on a freshly-deposited aluminum film for a constant sample voltage (with respect to the anode) of 5 V. Reprinted from Ref. 49 by courtesy of A. C. Nyce.

toward zero. The initial application of too large a positive sample potential causes breakdown of the initially thin oxide and the sample becomes the anode. This condition is highly undesirable as a strong local plasma will be formed over an area of the sample which will give rise to nonuniform anodization. Should a thick film be desired, the sample potential must be increased gradually so that it does not exceed the breakdown value. The most practical way to do this is to anodize at a constant current density; the sample potential will then become more positive with time in an approximately linear fashion. Anodization may then be terminated at a particular thickness or clamped at a predetermined value of sample potential while the current is allowed to decay. Figure 10 depicts the results of two constant-current anodizations carried out by Nyce [49] on pure aluminum films.

By application of a sample potential between the anode and metal film, oxide layers as thick as a few thousand angstroms have been formed on many materials. The ultimate thickness is determined by the space charge buildup in the oxide. This space charge buildup

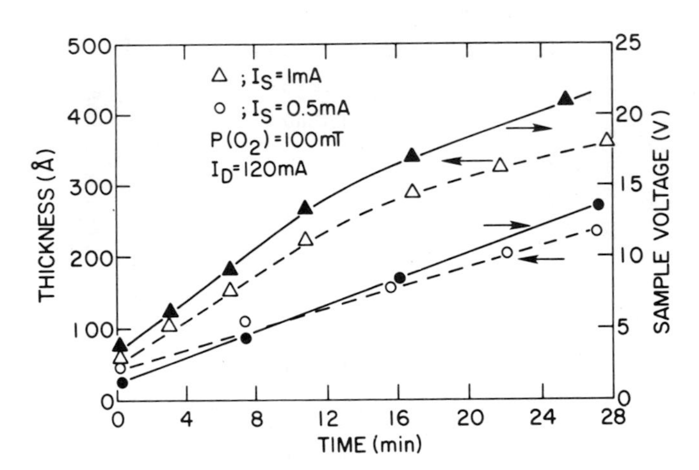

FIG. 10. The growth of Al_2O_3 on a freshly-deposited aluminum film for two values of constant sample current. Reprinted from Ref. 49 by courtesy of A. C. Nyce.

is dependent upon the material and its structure, the oxidation temperature, the nature of the oxide-plasma interface, and on any heat treatments given to the initial oxide layer before commencement of growth.

At this point it is instructive to define the faradaic or current efficiency η, a parameter which should be helpful in characterizing those anodization techniques in which the external sample current is separately observable. Using Faraday's law, the current efficiency η is defined by the equation

$$X = \frac{\eta M}{yF\rho A_s} \int_0^t i_s \, dt \tag{7}$$

where F is 9.65 x 10^4 coulombs, M is the gram molecular weight, ρ is the density, X is the film thickness, y is the valency of the metal ion, A_s is the sample area, and i_s is the sample current. For constant anodization, the overall efficiency may be expressed as

$$\eta(\%) = 1.608 \frac{y\rho}{M} \frac{A_s(\text{cm}^2)\ X(\text{Å})}{Q_s(\text{mA min})} \tag{8}$$

or for constant current anodization, as

$$\eta(\%) = 1.608 \frac{y\rho}{M} \frac{\frac{dX}{dt}\ (\text{Å/min})}{J_s\ (\text{mA/cm}^2)} \tag{9}$$

where Q_s is the total charge passed through the sample, and J_s is the sample current density; $y\rho/M$ has a value lying between 0.16 and 0.2 for most oxides. When all of the charge flow is necessary for the formation of oxide, this efficiency will have a value of 100%. Typical values of η for growth in the NG range from 0.5 to 5.0%. This is to be contrasted with values of η in the high 90% range for the anodization of most metals in aqueous solutions; an indication that in plasma anodization most of the charge flow stems from the copious supply of mobile electrons available in the NG.

b. Faraday Dark Space. Anodization proceeds essentially in a similar manner in the FDS and in the NG. Both Scharfe [43] and Copeland and Pappu [50] observed the growth rate to be somewhat less in the FDS. The data of Copeland and Pappu on the anodization of silicon in a cylindrically confined discharge indicate that the formation time at constant sample voltage is longer in the FDS than in the NG because of a smaller efficiency of anodization and not because of a smaller sample current. This is to be expected as both the average electron energy and density are less in the FDS than in the NG. Also, the electrons in the FDS are not as capable of generating negative ions either in the sheath or on the sample surface, as those in the NG.

c. Positive Column. Copeland and Pappu report that the anodization time in the positive column is less than in either the NG or the FDS. From their data, reproduced in Figure 11, it can be

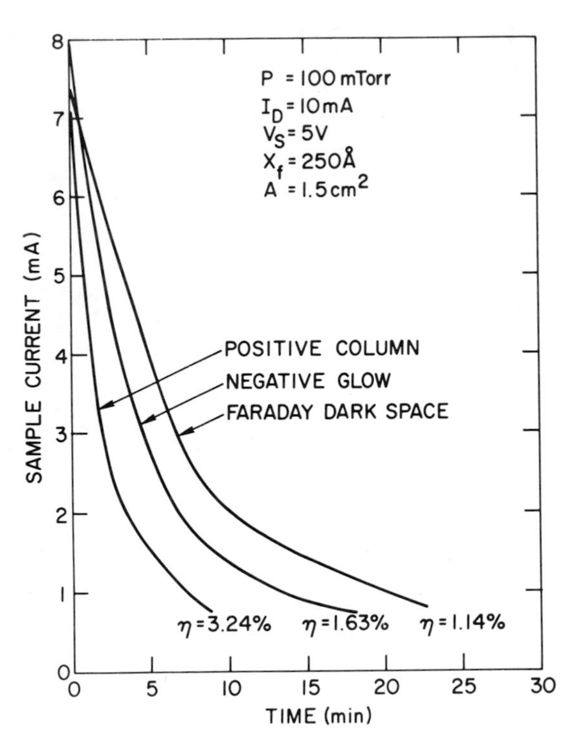

FIG. 11. Sample current decay during plasma anodization of a silicon film. Reprinted from Ref. 50 by courtesy of the American Institute of Physics.

seen that both the growth rate and the efficiency are about double the value obtained in the NG, and about three times the value observed in the FDS. Since the negative ion concentration is about two orders of magnitude greater in the PC than in the NG, anodic growth cannot be explained on the basis that extraction of negative ions from the plasma is the rate-limiting step [37]. Possible growth mechanisms will be discussed in Section V.

d. Anode Fall. Characterization of the work to be described here as growth in the "anode fall" may be misleading in that the anode referred to is not the main anode supporting the discharge, but rather the sample being anodized. The geometry in question, used by Jennings and McNeill [51] is depicted in Figure 12. The

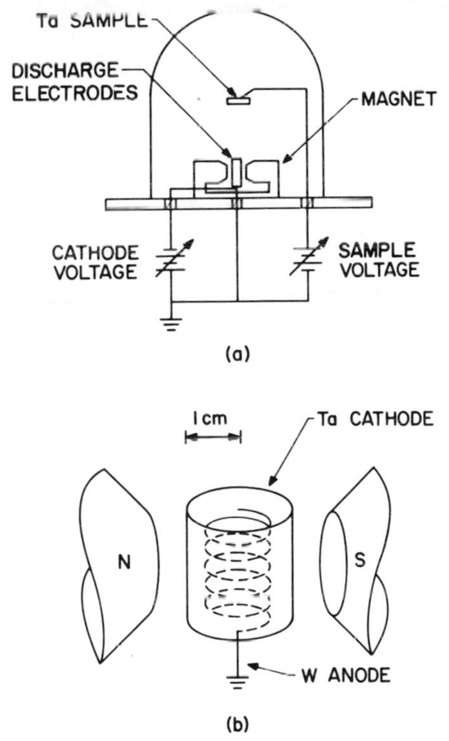

FIG. 12. Anodization apparatus used by Jennings and McNeill. (a) Vacuum system configuration. (b) Discharge electrode details. Reprinted from Ref. 51 by courtesy of the Electrochemical Society, Inc.

discharge anode and cathode are closely spaced and located in a magnetic field of 1,400 gauss. The magnetic field allows the electrons to make more ionizing collisions per unit volume resulting in a more intense plasma in a confined region. The sample is located outside the visible discharge produced by the main electrodes.

With this geometry anodization is taking place on an auxiliary anode, in the presence of an anode glow. As described earlier, the anode fall appears whenever the size of the anode is reduced; its presence is visibly indicated by the pale green glow immediately adjacent to the anode. This is confirmed in this geometry by two observations: first, Jennings and McNeill [52] observed the sputter removal of a gold film under positive bias in an oxygen plasma, a process that requires the presence of negative ions; second, the sputter yield decreases with increasing sample voltage, because oxygen plasmas become less electronegative at higher concentrations [12]. The O^- bombardment of the gold sample does not imply that this species is responsible for anodization; it only confirms the existence of an anode glow. The observed sputter removal of the gold film does not appear to affect oxide growth as the ion energies at the oxide surface are not likely to be greater than the oxide sputtering thresholds for the sample current densities under consideration.

The growth characteristics of typical anodizations done at constant current density are illustrated in Figure 13. The initial voltage jump at the onset of anodization was due to the sum of the potential drops across the nucleating oxide layer, the anode fall, and the rather weak plasma between the sample and the discharge electrodes. Growth was then linear until space charge accumulation began to reduce the internal field. The efficiencies quoted (17%) are in error; a correct value of 4.5% was obtained from this data [51] while efficiencies ranging from 0.5% to 8.0% may be derived from data given in later work [53,54].

The name "ion cathode" given by the authors to the discharge configuration seems inappropriate. The negative ions, whose role

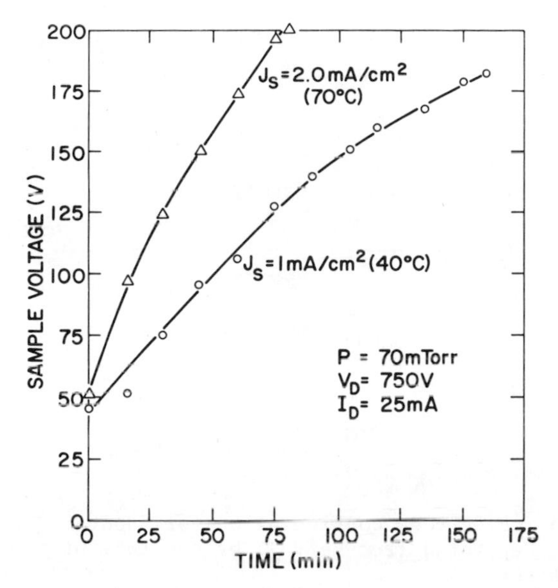

FIG. 13. Time-dependence of the sample voltage during growth of Ta_2O_5 at constant current density in the anode fall. After Jennings and McNeill; reprinted from Ref. 51 by courtesy of the Electrochemical Society, Inc.

was not clear, were actually generated in the anode glow which was observed in front of a small sample located more distant from the cathode than the NG. If the main discharge electrodes were in fact the source of the negative ions, then the plasma would have lost its electronegative character as the discharge current increased, and the anodization efficiency would have decreased, a conclusion which was not experimentally observed [53].

This technique is capable of rather efficient anodization of small samples at low (1-2 mA/cm^2) current densities. If the current density is made too large, the anode glow and oxide growth will not be uniform over the wafer surface, while the presence of too large a sample precludes the formation of an anode glow.

e. *Anode Glow*. The addition of an auxiliary anode near the sample has been used to produce high growth rates and efficiencies on niobium films [55]. The electrode configuration shown in Figure 14 has been used elsewhere [56] and has been modified only by the

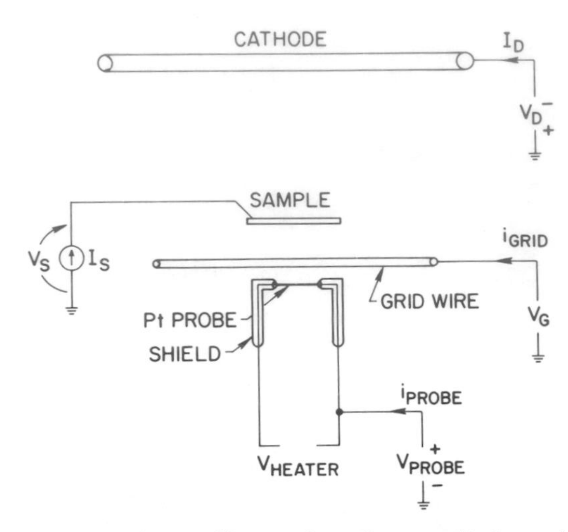

FIG. 14. Electrode configuration for anodizing with an auxiliary anode. Reprinted from Ref. 55 by courtesy of the American Institute of Physics.

addition of the auxiliary grid anode wire of 0.0025-cm-diameter platinum. By applying a positive bias between the grid and the main anode the plasma was grossly perturbed in several ways. Increasing the bias to 30 V made little change in either the electron temperature or the plasma potential; above 30 V, where the current in the grid wire was equal to or greater than the current to the main anode, both the plasma potential and the electron energy increased approximately linearly. The electron energy increased from 0.25 eV to 1.25 eV.

The effect of this change in the plasma environment on the faradaic efficiency is shown in Figure 15. A considerable increase in efficiency was seen at a grid bias of 50 V with efficiencies as high as 6-15% being realized as the region surrounding the sample was made to look more like the anode glow. This technique was also limited to small wafers as the efficiency decreased with large total sample currents.

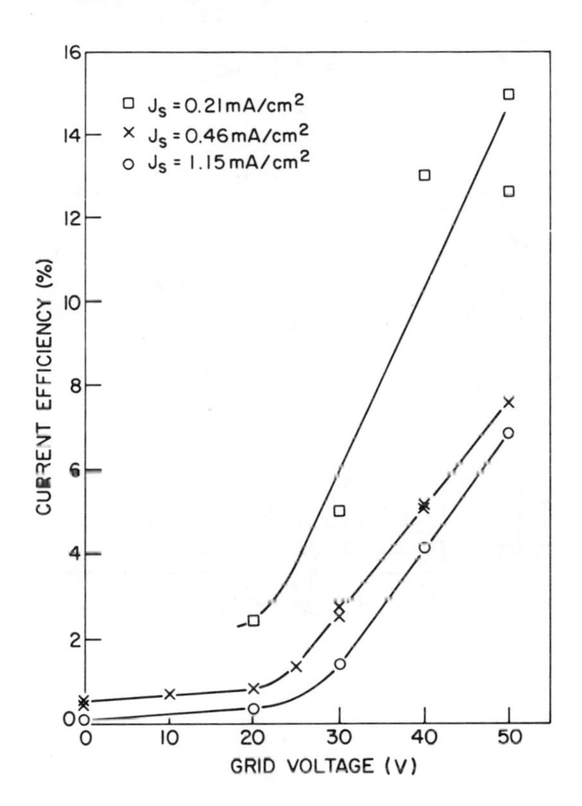

FIG. 15. Current efficiency for anodizing single-crystal niobium films at several anodizing current densities and grid voltages. Reprinted from Ref. 55 by courtesy of the American Institute of Physics.

B. Low Pressure DC Arc

Ligenza and Kuhn [9] developed a technique for anodization in a more intense DC arc. The hot hollow tantalum foil cathode developed by Ligenza and Povilonis [19] was protected by an argon sheath, while the anode and sample were in an oxygen atmosphere as depicted in Figure 16. Anodization proceeded in much the same manner as in the PC of a low pressure glow discharge except that the anodization rate was much faster (90 Å/min) because of the much higher electron and atomic oxygen concentration in the discharge. Anodization

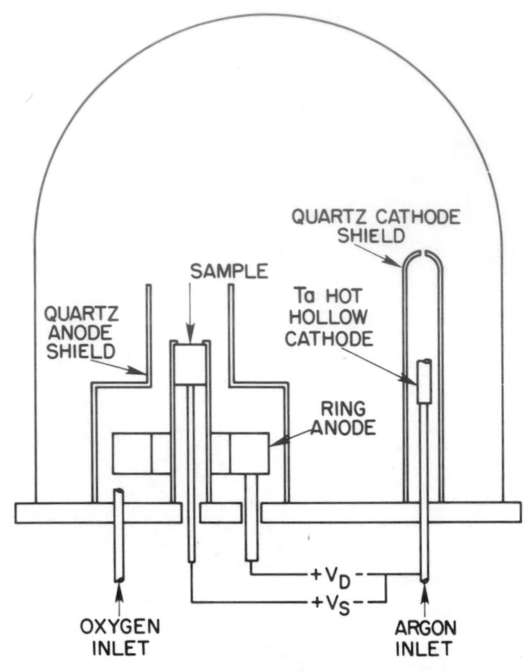

FIG. 16. Apparatus used for anodization in a DC oxygen arc. Reprinted from Ref. 9 by courtesy of the Cowan Publishing Corporation.

efficiencies were typically 0.7% for silicon in this geometry, as compared to 1.6% and 3.2% for the anodization of silicon on the NG and PC respectively [50]. Orcutt [57], using a similar apparatus for growing tantalum pentoxide, achieved efficiences in the range of 0.07% to 2.0%. This technique does not appear to involve a growth mechanism different from that observed in the low pressure glow discharge.

Orcutt [57] has shown that this technique cannot be easily scaled up to either faster anodization rates or multiwafer anodization. Increasing the sample current density to 100 mA/cm^2 produced little increase in anodization rate because the efficiency decreased; increasing the sample current beyond 100 mA/cm^2 resulted in the usual anode glow and nonuniform film growth. A first attempt at multiwafer processing showed uneven anodization of the samples.

C. Low Pressure AC Glow Discharge

1. RF Glow Discharge

Pulfrey et al. [58] and Pulfrey and Reche [26] have studied the anodization of silicon in a quartz tube (length = 100 cm; diameter = 5 cm) excited by an inductively coupled 1-MHz generator. Under normal operating conditions the oxygen pressure was 30 mTorr and the generator output power was 3 kw. Typical results of constant-current anodizations are shown in Figure 17. Films grown at lower current densities, e.g., 0.25-2.50 mA/cm^2, frequently were found to exhibit damaged regions where the oxide film was missing; irregular sputtering seemed to have taken place [58]. Films grown at current densities in the range of 5 to 30 mA/cm^2 were found to possess good electrical characteristics.

Mikhalkin and Odynets [23,24] and Makara et al. [25] were able to anodize tantalum and niobium in a 6.7-MHz RF discharge in the pressure range of 0.3 to 1.5 Torr. The growth rate and sample

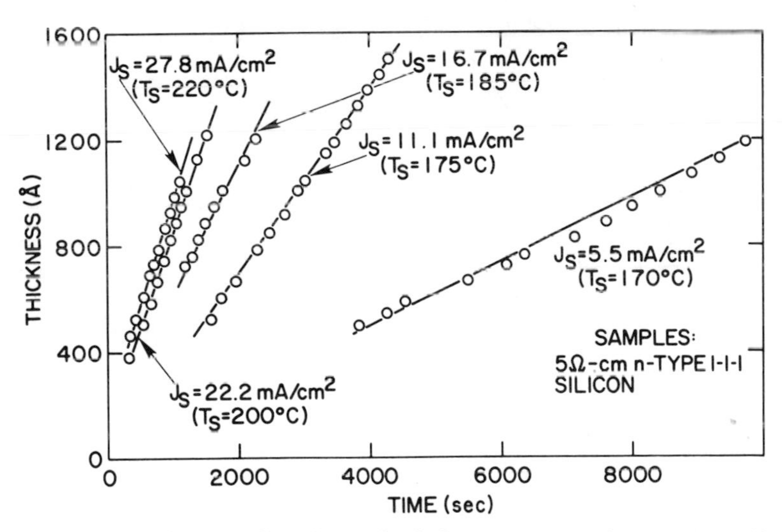

FIG. 17. The anodization of silicon samples in a 1-MHz discharge at P = 30 mTorr. Reprinted from Ref. 26 by courtesy of Pergamon Press.

voltage change for the anodization of niobium at 180°C are depicted in Figure 18. The efficiency of anodization was found to be very high (∿25%) for oxides under 1,000 Å, while it decreased somewhat at higher thicknesses. The enormous difference in efficiency observed in the two investigations [24,26] cannot be explained. The anodization of silicon also proceeds at a low efficiency in electrolyte solutions [59] so that a lower efficiency value is less unexpected for the plasma case.

2. Microwave Discharge

Ligenza [8] developed a technique for the growth of oxides in an oxygen discharge operating at 2.4 GHz. Ligenza's experimental system consisted of a quartz microwave cavity excited by a microwave power generator (Fig. 19). The cavity contained the usual sample and cathode counter-electrode, which together, acted much like a

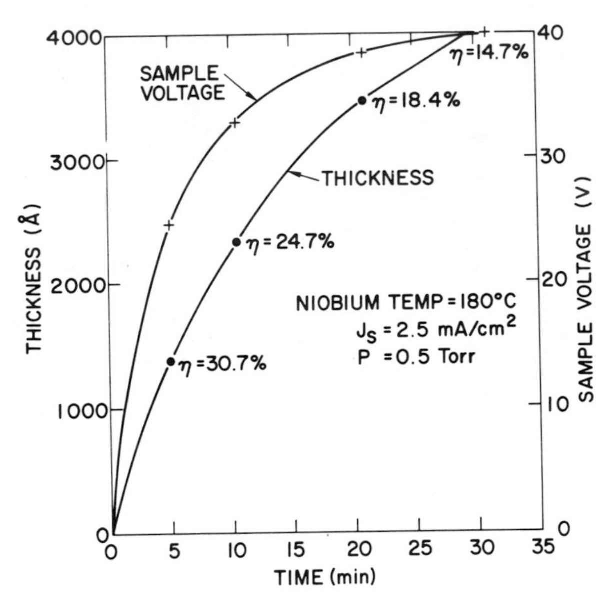

FIG. 18. Anodization of niobium in an RF discharge (P = 225 W, f = 6.7 MHz [25]. The sample voltage is measured between the sample and the cathodic counter-electrode.

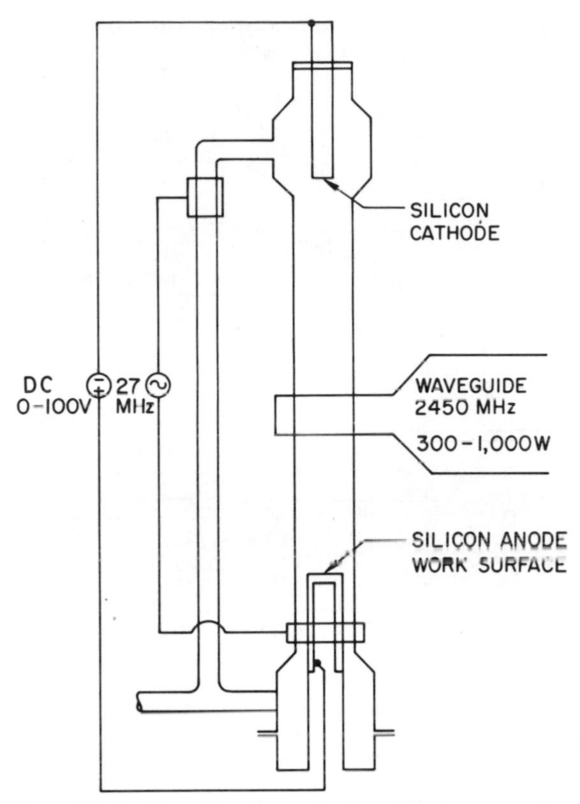

FIG. 19. Schematic diagram of the discharge tube and equipment for the observation of anodic oxidation of silicon in a microwave plasma. Placement of RF electrodes for sputter-cleaning the anode is also shown. Reprinted from Ref. 8 by courtesy of the American Institute of Physics.

floating double probe, as well as an RF generator for sputter cleaning the sample. The growth of silicon dioxide was observed with and without an applied DC sample voltage. The results of two oxidation runs, one with a sample voltage of 50 V and one with the sample floating are illustrated in Figure 20. The anodization rate was found to be independent of the sample voltage in the range 30 to 90 V; no rate-limiting thickness was found. Maximum growth rate was found to coincide with the presence of a striated region, containing an increased density of O_2^+, in the front of the sample.

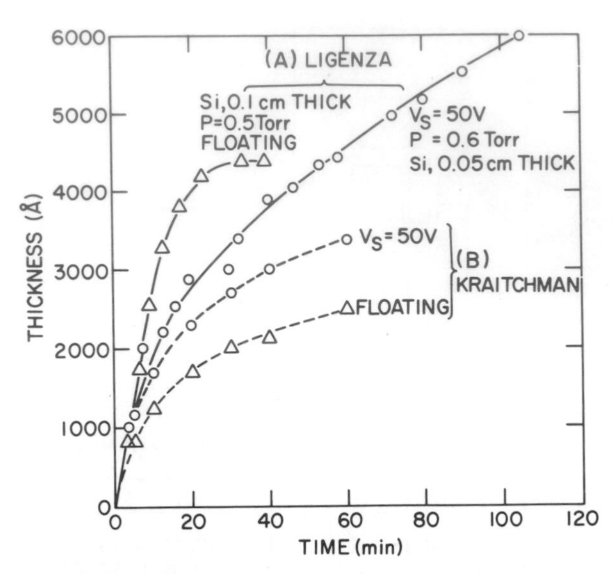

FIG. 20. Oxide growth in a microwave discharge with and without sample voltage. (A) Ligenza. (B) Kraitchman. Reprinted from Refs. 8 and 27 by courtesy of the American Institute of Physics.

Kraitchman [27] has pointed out that growth on a floating sample with a floating potential of -35 V with respect to the plasma [8] cannot occur by negative ion bombardment and that some other species is responsible. From the work of Ignatov [39] it is seen that O_2^+ bombardment is capable of producing appreciable growth rates in similar time periods. The application of a potential difference between the sample and the silicon counter-electrode immersed in a floating plasma can be regarded as a double probe [60]. Assuming the floating potential of the sample in Ligenza's experiment to be -35 V, the application of a 50-V potential difference causes the two electrodes to reference themselves with respect to the plasma in such a manner as to make the charge flow to the plasma equal at each electrode but opposite in sign. The more positive electrode will then be close to or perhaps a few volts positive with respect to the plasma, while the cathode will be approximately 50 V negative with respect to the plasma. Increasing the sample voltage to 90 V simply drives the cathode that much more negative while leaving the anode relatively unchanged; this is because the high mobility of the electrons is the controlling factor in determining how a double probe will align its potential with respect to a plasma. As the oxide grows, the voltage buildup across it is compensated by a movement of the potential both at the oxide surface and the cathode surface, toward the floating potential. Growth is seen to proceed parabolically with time and is faster with applied sample voltage than without it, the difference being presumably due to the presence of a small electric field in the oxide which accelerates the oxygen ions toward the silicon interface.

This technique, as many others, is limited to small sample areas, and to metal-oxide systems which are thermally stable at 400-500°C.

IV. EXPERIMENTAL PROBLEMS

In this section some problems relating to system contamination from the oxygen, chamber walls, and electrodes are described. Additionally, important complications arising from measurement of the anodizing voltage and thickness are discussed along with problems encountered in non-oxygen discharges.

Dell'Oca et al. [61] have discussed other experimental complications such as film uniformity, surface preparation, and electrical contact, along with difficulties involved in the interpretation of the so-called "anodization constant."

A. Gas Purity and Pressure

Gaseous impurities arise from outgassing of the chamber walls by energetic particles in the discharge as well as from the source gas. The most common contaminant in both cases is water vapor; the dew point of the source gas may be lowered by flowing the gas at reduced pressure through a liquid nitrogen trap. To minimize contamination from desorbed gases, the chamber may be vacuum-baked before the commencement of anodization and operated in a dynamic flow mode during anodization. However, Ligenza [8] was able to maintain a clean static discharge for as long as 1 hr after filling and flushing a few times in the presence of the discharge. A liquid nitrogen cold finger in the work chamber is useful in pumping desorbed water vapor. Because electrons and ions are so proficient at cracking or polymerizing and desorbing vapors, good vacuum practice is in order.

Water vapor contamination has been shown to increase the thickness of thin (<50 Å) oxides grown on electrically floating electrodes by as much as a factor of two. Huber et al. [62], Pollack and Morris [48], and Miles and Smith [6] all report increased thicknesses when no liquid nitrogen trap was used. This may be due either to the increased generation of atomic oxygen in the presence of hydrogen as discussed previously [18,31], or to the OH incorporation in the film.

Paradoxically, thicker oxides grown with an applied sample potential have always shown decreased growth rates when no liquid nitrogen trap was used [61,56], or when water vapor was deliberately added [28,63]. Dissociative ionization of water vapor results in a lower concentration of O_2^+ in the discharge because the ionization energy of hydrogen is less than that of oxygen. Hydrogen also effectively lowers the temperature of the electrons in the discharge.

Oxidation rates have been observed to be pressure dependent. O'Hanlon [56] has shown that the growth rate in the NG was a maximum in the pressure range of 50 to 70 mTorr. Orcutt [57] has observed a similar behavior with the DC arc, while Ligenza found that growth rates in the microwave discharge were greatest at 0.5 Torr. In the case of the DC glow discharge the pressure at which maximum growth took place was dependent upon the discharge current and cathode-sample spacing; maximum growth occurred when the electron concentration at the sample location was greatest [56].

B. Electrode Sputtering

In DC glow discharges, cathode sputtering has been directly observed in some work while it has been postulated to be the cause of linear growth rates at constant voltage in other work [61,64]. The amount of sputtering is a function of the discharge current, cathode fall potential, cathode material, cathode-sample spacing, and relative orientation. In general, sputter contamination of the sample is increased at higher discharge currents and voltages, and smaller cathode-sample spacings. Using an 80-mA 800-V discharge, sputter deposition rates in the range of 0.5 to <0.02 Å/min have been observed for an aluminum ring cathode as the sample position was changed in the range of 5 to 20 cm from the cathode [64]; however, Leslie and Knorr [65] observed no sputtering (<0.005 Å/min) at a distance of 8 cm from a 40 cm^2 planar aluminum cathode operating at a discharge current of 10 mA in the same pressure range (50-70 mTorr). Since the quantitative results are very much dependent upon experimental conditions, such as water vapor partial pressure, a wafer must be checked by a technique such as in-situ ellipsometry [65] or X-ray fluorescence [64] to determine the importance of sputter contamination for the system of interest. In any system in which the probe floating potential is observed to drift with time, sputter contamination must be considered. The present author has also observed temporal floating potential changes whenever organic insulators such as nylon or Teflon were used to isolate current-carrying wires from the discharge. Consequently, the sample holder used by the author was patterned after the excellent design of Jennings et al. [51].

C. Measurement of Oxide Surface Potential

In order to correctly estimate the average electric field in the oxide, both the thickness and the potential drop across the oxidizing sample must be known. The real problem is relating the measured quantity to the potential at the surface of the oxide facing the plasma. In early work [6,66] the total applied sample voltage was considered to be dropped across the oxide, implying that the oxide surface potential was at the anode, or ground potential, whereas in reality, the sample behaved as a large-area Langmuir probe. Several authors have pointed out the inadequacy of this method and have suggested various approaches to solving the problem. Jackson [67] and Scharfe [43] have placed small probes close to the sample, and have considered the oxide potential drop to be the difference measured between the sample and the plasma potential of the probe. Olive et al. [68] have formulated the potential drop to be the potential between the sample and probe plus the difference in work functions between the probe and samples. Some of the problems with this method can be avoided by taking an i-v characteristic of the sample with respect to an invariant probe at some sufficiently close distance to the surface for various oxide thicknesses. If the sheath profile is dependent only on current density then a reasonably approximate incremental field can be obtained in this manner for a constant current density.

O'Hanlon [69] has suggested that the potential drop across the oxide at the conclusion of constant voltage anodization is approximately the difference between the sample voltage and the plasma floating potential. Growth was observed with zero sample voltage. Ramasubramanian [70] has observed growth in situations where the sample voltage was less than the plasma potential, but greater than the floating potential, a result which supported this model. Two phenomena will affect the accuracy of this approach: the dependency of the floating potential on material [70], and the amount of bound charge in the oxide. Nyce [49] has developed an alternative approach and formulated the average oxide drops to be

$$V_{ox} = V_s - V_p(i) + (V_{fg} - V_{fs}) \tag{10}$$

where V_{ox} is the oxide potential, V_s is the applied sample potential, $V_p(i)$ represents the potential of a geometrically identical gold probe with the same charge flow as the sample, and $(V_{fg} - V_{fs})$ is the difference between the floating potentials of the gold probe and the sample. Again, assuming a sheath profile that depends only on current density, this model makes an approximate correction for work function differences between probe and sample as well as space charge in the oxide. In all of these methods, the electrical double layer at the electrodes and the free energy of formation of the oxide have been largely ignored [68].

Considering the crudeness of all of these models, the gross manner in which the sample perturbs the plasma, and the space charge in the oxide, it is surprising that the scatter in the Tafel plots is as small as has been observed [63,25].

D. Film Thickness Measurements

Tolansky interferometry is a direct method that can be used to measure the thickness of etched and silvered anodic oxides [69], and when combined with the measurement of the step height of the oxide over an unanodized portion of the metal, yields the density. The disadvantage of the direct technique is obvious: the time-dependence of film growth cannot be followed on a single sample.

Several indirect thickness measurement methods can also be used on the completion of anodization. Tibol and Hull [66], Miles and Smith [6], and Jennings and McNeill [51] measured the capacitance of the films and assumed a dielectric constant in order to calculate their thickness, while Miles and Smith [6] deduced their thickness by comparing the tunnel current in thin film samples with an appropriate theoretical model. More common has been the use of a set of pre-anodized step gauges [9,43,67,69]; this is a simple accurate method provided that the index of refraction of the film is checked independently by liquid immersion.

Oxygen weight gain was used by Nazarova [40,41] to calculate the in-situ oxide thickness; he measured both the weight gained while the sample was suspended on a quartz-beam microbalance and the pressure loss in the oxidation cell by means of a McLeod gauge. For this method to be accurate, anodization must be performed immediately following several glow discharge cleanings and flushings. Vrba and Woods [71] have anodized aluminum films evaporated on a quartz crystal used in a film thickness monitor. This technique requires polished crystal surfaces in order to obtain films of known area, and unless the frequency-generating electronics is isolated electrically from the glow discharge anode, the sample is electrically at the anode potential. One is then limited to the study of constant voltage growth kinetics at one applied potential.

Ellipsometry has been extensively applied to gas discharge anodization by Locker and Skolnick [72], Nyce [49], Lee et al. [63], and Leslie and Knorr [65], to study the growth kinetics of several metal oxides. With the exception of the latter work in which the data were automatically recorded every 5 sec, all others used manually-operated equipment. The beauty of this method is that it yields the index of refraction and thickness simultaneously for each measurement, provided that the refractive index of the bare metal substrate is known. Table 1 lists the refractive indices of several substrates on which oxide growth has been monitored extensively by in-situ ellipsometry.

If one assumes that the oxide has a single-valued refractive index that is lossless, then mathematically one has a unique solution, that is, two knowns, the polarizer and analyzer angles (or ψ and Δ), and two unknowns, the index and thickness. However, if the film is modeled to have two discrete layers of differing index or

TABLE 1

Refractive Indices of Several Bare Substrate Materials

Metal	Refractive index	Wavelength, Å	Reference
Ta	3.3-i2.3	5461	59
Ta	2.3-i2.6	6328	65
Nb	3.6-i3.6	5461	59
Nb	3.0-i3.6	6328	73
Si	3.861-i0.017	6328	74
Al	1.63-i7.54	6328	49
Au	0.1975-i3.19	6328	49
GaAs	3.923-i0.304	5461	75

one index with a loss component, then the calculation for a single datum is not unique. The ambiguity can be removed by taking a series of points while growing the film to sufficient thickness such that the ellipsometric ψ-Δ curve completes more than one cycle. In this manner differences of interpretation such as the two-layer model of the refractive index of tantalum by Lee et al. [63] which has been seen in tantalum oxide formed in dilute acid [76], and the model of a single-valued but absorbtive index by Knorr and Leslie [73] can be resolved.

The s-light reflectivity technique has been used by Pulfrey and Reche [26] to effectively measure the sample thickness in situ. To implement this method, one records the magnitude of the s-light (electric vector perpendicular to the plane of incidence) that is reflected from the growing oxide. Upon completion of oxidation, the refractive index must be measured, in order that the reflectivity versus time data can be compared to reflectivity versus thickness data for the appropriate refractive index; from this a plot of thickness versus time can be constructed.

E. Non-oxide Films

There has been only limited success in growing anodic films in discharges other than oxygen. Lewicki and Mead [77,78], Lewicki and Maserjian [79], and Uemura et al. [80] have subjected freshly evaporated and electrically floating aluminum films to a nitrogen glow discharge to produce aluminum nitride films in the thickness range of 30 to 140 Å.

The nitriding process was subjected to the criticisms that trace amounts of water vapor produced oxide on the freshly evaporated surface, and that the aluminum nitride was really reactively sputtered from the aluminum cathode [64]. Uemura et al. have demonstrated that aluminum nitride films were formed when the partial pressure of water vapor was less than 10^{-8} Torr. They also confirmed that aluminum nitride films were formed by both direct nitriding and reactive sputtering simultaneously; aluminum nitride was observed to form in the presence of a tantalum cathode, and it was also observed on a glass slide with an aluminum cathode. Both Freiser [81] and the present author were unsuccessful in growing oxides thick enough to have visible interference patterns on silicon samples that were biased positive with respect to the anode. The author also has attempted anodic formation of carbides and sulfides. Several metal films immersed in a methane plasma became coated with a rapidly-growing high-resistivity carbon film, while no visible film grew on a lead foil immersed in a heated sulfur plasma.

The only report of cathodic film formation in a non-oxygen discharge is that of Schmellenmeier [82] who was able to grow tungsten carbide on a tungsten cathode supporting an acetylene discharge.

V. OXIDATION RATE-CONTROLLING MECHANISMS

A. Introduction

Several growth-rate-limiting mechanisms have been proposed to explain the observed thickness-time behavior of plasma-grown oxides. Closer examination suggests that the problem is many faceted with the rate-limiting mechanism depending on the type of discharge and the nature and structure of the metal being anodized. The growth rate can be limited by factors external to the sample such as the influx of electrons or some oxygen species, or an internal mechanism such as the material, its structure, or the local electric field. In fact the rate-limiting mechanism may change during the course of oxide growth. This section represents an attempt to illustrate situations in which different rate-limiting mechanisms have been observed.

B. External Mechanisms

Because the several discharges used in anodization differ so greatly, the effect of each discharge will be discussed separately.

1. Microwave Discharge

Microwave oxidation is limited to materials that do not undergo thermal oxidation at the ion temperature of the discharge, typically 400-500°C. Thorough experimental work has been done only on silicon, making comparisons with other techniques somewhat less than typical, because of the uniquely low current efficiencies obtained on silicon with all techniques. Analysis of the growth behavior in a microwave discharge reveals that the oxidation is a diffusion-limited process [8,27]. At the oxidation temperatures of interest only a high concentration gradient would appear to explain the observed growth rates. This concentration gradient could be provided by a large flux of some excited oxygen species such as O_2^+ or O^+. It appears that the rapid oxidation rate is not due to negative ions, as none would strike an electrically floating sample with its large negative bias [27]. Kraitchman [27] has also observed that the intense bombardment by excited oxygen causes film sputtering, and has modeled the growth rate as

$$\frac{dx}{dt} = \frac{k}{2x} - s \tag{11}$$

where k is the parabolic rate constant, and s is the sputtering rate constant. This implies that the oxide asymptotically approaches a rate-limiting thickness of $x = k/2s$. In other words, in the thick film limit, a large energetic flux again limits the film growth, but by a combination of mechanisms.

2. Low Pressure Glow Discharge

The role played by gas-phase negative ions in anodization has been argued at some length. The weight of the evidence shows that the growth rate of plasma oxides is not limited by the diffusion of negative ions from the plasma to the surface. The maximum negative ion concentration in the NG has been shown to be of the order of $10^8/cm^3$, implying for most metals an oxide growth rate of ∿0.3 Å/min. This contrasts with observed rates of order 3-30 Å/min, and even as high as 800 Å/min [49]; clearly, the extraction of oxygen anions from the plasma is not an adequate explanation for the observed rates. The FDS and PC which are quite different in their particle concentrations show only factors of 2 or 3 difference in growth rate. Olive et al. [83], using a combination of DC and RF voltages on a sample undergoing oxidation, have shown that large changes in the sheath potential affected the growth rate only slightly as compared with the expected change if growth were dependent upon the supply of negative ions from the plasma.

Alternatively, ionization in the space charge sheath and surface ionization of adsorbed atomic oxygen have been considered. One of these two mechanisms would have to be invoked to explain the drastic effect of the extra grid on the anodization efficiency in the low pressure glow [55].

3. Low Pressure DC Arc

Growth in the DC arc is similar to growth in the low pressure glow in that the current efficiency is similar in both cases. However, the discharge concentration and the anodization rate are greater in the arc than in the glow [9]. The reported high rate is due in part to the high temperature used (225°C); data at other temperatures were not given [9]. There is the possibility that the rate of entry into the oxide of electrons from the plasma limits the field which may be set up in the oxide, which in turn limits the rate of growth; e.g., the high growth rates in this plasma may be due to the relatively large electron and atomic oxygen concentrations [63].

C. Internal Mechanisms

1. Electric-field-limited Growth

Vrba [71] demonstrated that the initial oxidation phase of aluminum is controlled by the constant current (and constant electric field) giving rise to a linear growth rate. As the oxide potential drop increased to the maximum value permitted by the external circuit ($V_{anode} - V_{surface}$), growth was described by the generalized Mott-Cabrera law. Logarithmic growth was found to fit for a period of time, after which growth proceeded slightly faster. This increased growth rate is due to the fact that the oxide potential drop is really not constant owing to the drift of the oxide surface potenrial toward the floating potential.

Lee et al. [63] and Mikhalkin and Odynets [24] have shown that $J_i \simeq A \exp BE$, while the electron current obeys a Poole-Frenkel dependence $J_e \simeq A' \exp B'\sqrt{E}$ (see Fig. 21). Thus the field in the oxide determines the maximum ratio of J_-/J_e, or current efficiency; the ion current may be less if limited by surface processes, etc. In most systems studied, this efficiency is quite low because of the large supply of mobile electrons.

Nyce [49] has shown that for a given thickness and anodizing current, on changing the structure of Al_2O_3 from an amorphous to a polycrystalline form, the electric field increased, giving rise to an increased ration of J_-/J_e. In this case the current efficiency and growth rate were observed to be as high as 55% and 800 Å/min, respectively. These increases were attributed to both the increase in electric field, and to an increase in ion current along grain boundaries. The heat treatment was carried out by heating the 60-Å

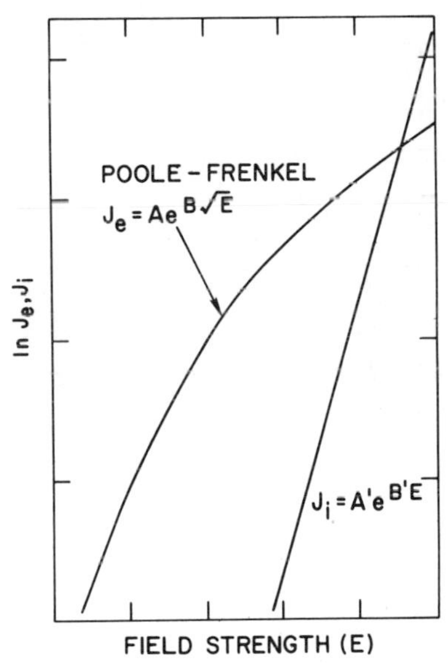

FIG. 21. Qualitative dependence of the electronic (J_e) and ionic (J_i) components of the anodizing current. Reprinted from Ref. 49 by courtesy of A. C. Nyce.

nucleating layer of oxide to 200°C for 1 hr, followed by anodization at 25°C. Growth rates for the amorphous and low-temperature polycrystalline oxides are shown in Figure 22.

2. Space-charge-limited Growth

Norris and Zaininger [84] and Waxman and Zaininger [85] have studied charge trapping in metal-Al_2O_3-Si transistors formed by plasma anodization. They concluded that the positive and negative oxide charges were due to the trapping of holes and electrons which were injected into the oxide during formation, similar to the behavior sketched in Figure 23(A). From this model one would expect a C-V measurement to indicate a net positive oxide charge as outlined in Figure 23(B). According to this charge distribution, the positive charge near the silicon should and did produce depletion (inversion) in p-type substrates, and accumulation in n-type substrates.

Furthermore this trapped charge, which is accumulating during anodization, modifies the linear electric field in the oxide in

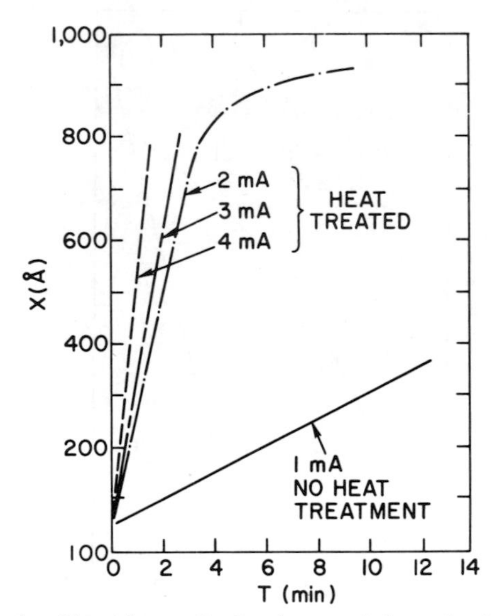

FIG. 22. Anodization of aluminum with and without heat treatment of the 60-Å initial oxide. Reprinted from Ref. 49 by courtesy of A. C. Nyce.

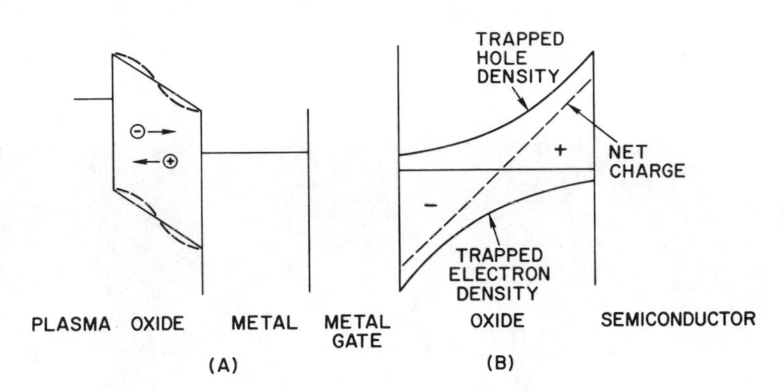

FIG. 23. (A) Potential energy distribution in the Al-Si system during anodization of the aluminum. (B) Trapped charge distribution in the Al_2O_3-Si system after anodization of the aluminum. Reprinted from Ref. 84 by courtesy of the United States Department of the Navy.

such a way as to decrease the field at either surface of the oxide and increase it in the center; see dotted curve in Figure 23(A). Regardless of which ion (metal or oxygen) is mobile in the oxide, the trapped space charge will cause the anodization rate to decrease as the oxide thickness increases. This is most likely the cause of the initial deviation from linearity in the constant-current growth mode; see, for example, Figures 10 and 13.

3. Stress-limited Growth

Both Pulfrey and Reche [26], and Morgan [86] have observed breakdown in SiO_2 and Al_2O_3 films, respectively. Their evidence pointed to breakdown due to large compressive stresses in the oxide leading to stress relief by visible deformation. This deformation, shown in Figure 24, takes the form of blisters; some of the blisters have collapsed after removal of the anodizing current indicating some stress relief. Pulfrey and Reche further observed that a current density-thickness product seems to bear some relation to the film growth limit. For SiO_2, they find thicknesses greater than 3,000 Å possible for $J = 2$ mA/cm^2, while breakdown occurs at 1,200 Å for $J = 30$ mA/cm^2.

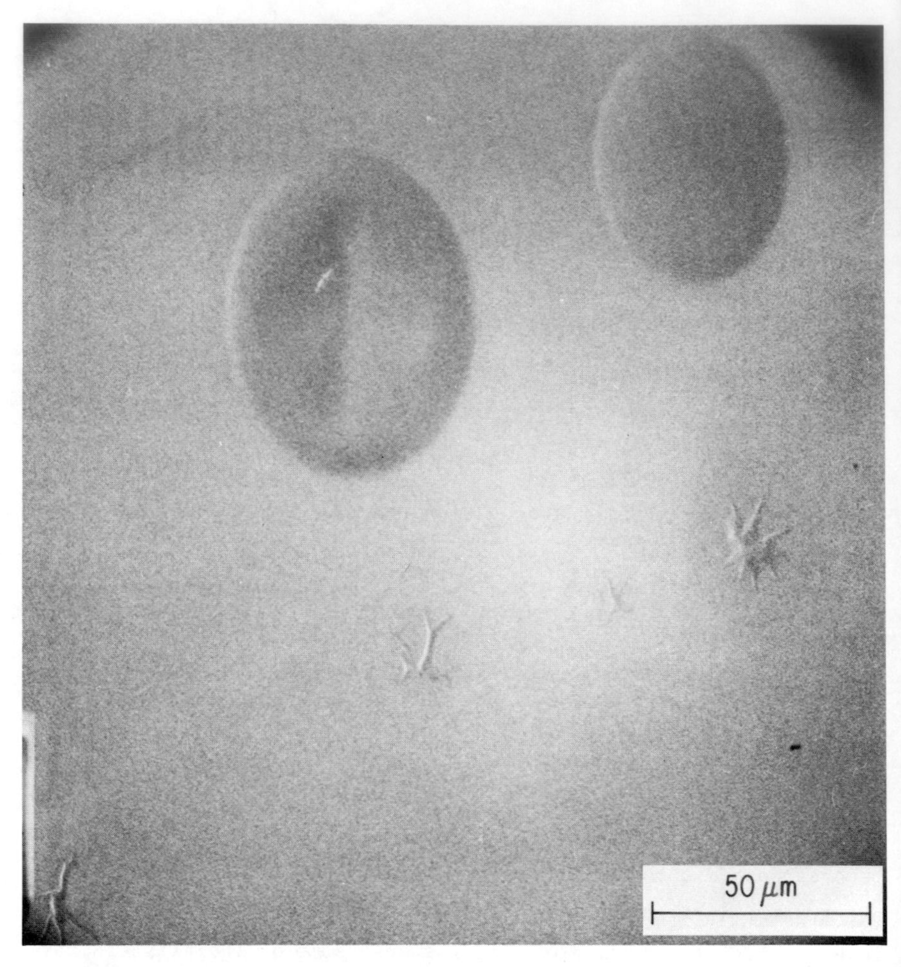

FIG. 24. Blisters caused by the compressive stresses in plasma-grown SiO_2 on silicon. Reprinted from Ref. 26 by courtesy of Pergamon Press.

One should note here that temperature is an important parameter in the last two rate-limiting steps discussed above, namely, the space-charge-limited growth, and the stress-limited growth. For example, during anodization in microwave plasmas at 400-500°C [8,27,28], no such limits were seen. This temperature was greater than the temperature used (350°C) to anneal out the bound charge in MOS structures, and it undoubtedly enabled considerable stress relief.

VI. OXIDE AND NITRIDE FILM PROPERTIES

Relatively few of the many plasma anodization papers report systematic data on the physical, optical, and electrical properties of the films formed. Some contain meaningless data such as the resistance or capacitance of films of unmeasured thickness, while the majority of them contain no such data at all. In this section the material- and device-related properties of films formed in oxygen and nitrogen discharges are summarized. At the end of this section, a table classifying published papers by material and discharge type is presented. Absent from this table (Table 2) are papers or chapters of a review nature only.

A. Aluminum Nitride

Aluminum nitride films have been formed by a combination of DC reactive sputtering and direct nitridation [77,78,79,80]. The energy-wave vector relationship has been determined over much of the forbidden band [77,78], and the current-voltage characteristics have been studied [77,79]. The Al-AlN energy barrier was found to be 1.76 eV while the AlN-Mg counter-electrode barrier was found to be 1.64 eV [78]. Uemura et al. [80] have determined the films to be of the wurtzite structure, as is bulk aluminum nitride, with a grain size of 50 Å; the (001) plane was found to be parallel to the surface. Voltage-controlled negative resistance was observed in Al-AlN-Au structures.

B. Aluminum Oxide

The metal-oxide barrier heights of thin plasma-anodized Al-Al_2O_3-Al sandwiches have been measured by several techniques including the fractional current increase with temperature [48,87,88], logarithmic conductivity [89], and photoemission [90,91]. There appears to be a greater variation in the plasma-anodized barrier height data (1.65-2.6 eV) than in barrier heights between plasma and thermally oxidized structures. Apart from any differences arising from the preparation of samples, Lewicki and Mead [92] and Gundlach and Kadlec [93] have shown that the barrier height increases with thickness in the range of 25 to 90 Å. This effect combined with the increase in barrier height seen under 3.7 eV illumination clearly demonstrates that the space charge in electron traps is responsible for the modulation of the barrier height [93].

Photoemission studies by Goodman [91] have shown that both electrons and holes can be photoemitted from aluminum into annealed plasma-anodized Al_2O_3. From this work the bandgap was determined to be 5.1 eV. This compares with the value of 5.2 eV obtained from optical absorption by Miyazaki et al. [94]. The optical bandgap of wet anodized Al_2O_3 appears to be 6-7 eV, while that of crystalline Al_2O_3 is 8-9 eV. The origin of these differences is not clear.

The crystal structure of anodic Al_2O_3 formed by various plasma techniques has been shown to be amorphous by electron diffraction [3,4,39,49,95] and infrared absorption [96] studies. Ignatov [39] has reported weak lines due to the crystalline phase γ'-Al_2O_3 (cubic lattice constant 3.95 Å) in anodic films after long anodization times on the anode of a DC cell. Ignatov [4,39] has studied the growth of Al_2O_3 on the cathode and found it to be entirely polycrystalline γ'-Al_2O_3; however, subsequent to the formation of the oxide to a thickness of $\sim$300-400 Å, the film became pitted by sputtering. The index of refraction of the amorphous oxide was variously found to be 1.67-1.70 [95], 1.75 [97], 1.639 [49]; Nyce [49] found the index of refraction of heat-treated polycrystalline alumina to be 1.71. Dielectric constants reported ranged from

6 to 8.5 [98], and from 8 to 8.5 [99]. Norris and Zaininger [84] obtained a dielectric constant of 7.6 and film densities in the range of 3.1 to 3.3 g/cm^3, the latter being in agreement with the published value for wet anodic Al_2O_3 viz. 3.1 ± 0.2 g/cm^3 [100].

Terry and Komarek [101] found the breakdown strength to be 8 x 10^6 V/cm, while Nicol [102] found the breakdown strength to obey a $t^{-1/4}$ power dependence with E_b = 6 x 10^6 V/cm at 100 Å. A detailed examination of the experimental results of the two papers [101,102] shows that the films were at least partially, if not mostly, formed by DC sputtering of the aluminum cathode.

Several properties of aluminum oxide have been measured during the course of device studies. Tibol and Kaufman [98] have fabricated capacitors of Al-Al_2O_3-Al sandwiches with capacitances of 0.2 μF/in.^{2} (50 V anodizing potential). The temperature coefficient of capacitance was found to be 340 ppm/°C (-65°C to +150°C); the dissipation factor was 0.5 at 100 kHz. Norris [99] obtained a dissipation factor of 0.007-0.01 in the range of 100 Hz to 500 kHz, compared with 0.004-0.01 in the same range for vapor-deposited Al_2O_3 [103].

The density of surface states of the gate-semiconductor interface has been studied in Al_2O_3-Si insulated-gate field-effect transistor (IGFET) structures [95] and Al_2O_3-CdSe thin-film transistor (TFT) structures [104]. In both cases the samples possessed a high density of positive oxide charge (10^{12}/cm^2) after fabrication, whereas on annealing in hydrogen or forming gas at temperatures of 300-400°C for 1 hr the oxide charge density was reduced to around 2 x 10^{10}/cm^2. The TFT devices fabricated in this manner were found to be far more stable than those fabricated with evaporated SiO, while the IGFETs were equal to those made with the best quality thermal SiO_2. Pappu and Boothroyd [105] have fabricated metal-plasma anodized Al_2O_3-SiO_2-semiconductor (MAOS) capacitors; however, insufficient data are given for a detailed comparison with other fabrication methods.

The radiation resistance of Al_2O_3-MOS devices has been extensively reported [95,106,107]. When these devices were subjected to

1-MeV electron bombardment of various fluence levels and bombardment-bias conditions, they showed an improvement in radiation resistance more than an order of magnitude over the conventional SiO_2 or chemical vapor-deposited Al_2O_3 units. At up to 10^{13} e/cm^2 under either polarity bias, no oxide charge buildup or interface charge generation was detectable. The principal improvement was found to be for positive gate bias where the SiO_2 is particularly vulnerable.

In spite of the excellent electrical properties of the Al_2O_3-Si interface, active device fabrication progressed slowly. A major obstacle was the anomalous etching property of plasma-grown Al_2O_3 on silicon when exposed to either hot phosphoric acid or buffered hydrofluoric acid. Micheletti et al. [96] reported a tensile stress of 10^{10} $dynes/cm^2$ in Al_2O_3 on silicon, after anodization was completed; no appreciable stress was observed in the deposited aluminum on silicon starting material. No anomalous etching was found in chemical vapor-deposited Al_2O_3 films with the same stress. It is possible that the large anodization current caused a contraction in the film due to electrostriction which affected the adhesion of the oxide to the silicon, and that the etchant penetrated to the interface region via pinholes or grain boundaries. Using scanning electron microscopy, Pulfrey and Reche [26] and Morgan [86] have observed blisters in Al_2O_3 films. In Pulfrey and Reche's work the blisters appeared only in samples anodized at high current densities ($\sim$10 mA/cm^2).

C. Copper

Copper was oxidized on both the anode and cathode of a DC glow discharge. It was determined by electron diffraction to be Cu_2O when formed on the anode and CuO when formed on the cathode [41].

D. Gallium Arsenide

The oxide formed on GaAs in a microwave plasma was found to be completely amorphous, with an index of refraction of 1.78, a dielectric constant of 3.9, and a breakdown strength of 5×10^6 V/cm, for films of thickness 3000-4000 Å. In contrast to the thermal oxide, the plasma oxide was found to be soluble in hot water [29]. The oxide formed in a DC low pressure glow at 30 Å/V has an index of refraction of 1.82-1.87 [75].

E. Gallium Arsenide-Phosphide

Sugano and Mori [108] have studied the oxidation of $GaAs_{1-x}P_x$ (x = 0 to 1) in a high frequency glow discharge. The films were observed to have a complex composition, including polycrystalline β-Ga_2O_3, $GaAsO_4$, and $GaPO_4$. The refractive index was found to vary from that of β-Ga_2O_3, at the exposed surface, to that of the substrate, at the substrate-oxide interface. The etch rate was found to vary considerably with the arsenic-phosphorous ratio.

F. Garnets

Simpson and Lucas [109] anodized films of various controlled yttrium-iron and gadolinium-iron alloy compositions. The iron-yttrium stoichiometry of the FeY and 5Fe3Y alloys was reproduced in the oxide. A small amount of aluminum sputter contamination was observed. All oxides were found to be completely amorphous.

G. Germanium

O'Hanlon [69] has oxidized germanium to form water-insoluble GeO_2. The films were found to have a refractive index of 1.67, a dielectric constant of 6.4, and infrared absorption peaks at 850 and 650 wave numbers.

H. Hafnium

In a preliminary attempt at fabricating HfO_2 gate oxides for IGFETs, Norris and Zaininger [84] found the index of refraction to be 2.0, and the density to be 7.85 g/cm^3, compared with the bulk density of 9.68 g/cm^3.

I. Lead

The superconducting energy gap of PbO was found to be $\Delta_0 = 1.1 \times 10^{-3}$ eV [110] and $\Delta_{4.2} = 1.2 \times 10^{-3}$ eV [47], for growth of a thin oxide on an electrically floating sample in a DC low pressure discharge.

J. Nickel

Ignatov [39] found that the oxide formed on either the cathode or the anode was crystalline with the NiO composition; the 400-Å nickel film could not be completely oxidized.

K. Niobium

The index of refraction of niobium oxide has been reported to be 2.30 [73], and 2.0 [54]; the two-layer model of Lee et al. [63] gave 2.15 and 2.37 for the inner and outer layers, respectively. Dielectric constant values ranging from 18 to 43 have been reported [25,54,110]; the resistivity was reported to be 9×10^{15} Ω cm [54]. The superconducting energy gap at 0°K (Δ_0) was 1.5×10^{-3} eV [110]. The Nb-Nb_2O_5-Bi structures prepared by the present author performed as bistable switches with properties similar to those of the annealed wet-anodized devices [111].

L. Silicon

The composition of the oxide formed by plasma anodization was shown to be stoichiometric SiO_2 [8,26] with a uniformly high breakdown strength of 0.7 to 1.2×10^6 V/cm [8,26,44,112] for films over 1,000 Å thick, regardless of the experimental technique used to produce the oxide. The dielectric constant was found to vary between 3.5 and 4.0 with the value 3.9 equal to that of the thermal oxide, routinely attainable [8,9,26,58,112]. A resistivity of 10^{16} Ω cm was typically obtained [8,26,44]. The index of refraction was typically 1.455-1.47 [8,26,28] as compared with 1.466 for steam-grown oxides and 1.463 for oxides grown in dry oxygen [28].

The etch rate reported by Skelt and Howels [28] was 2.58 Å/min for plasma SiO_2 in p-etch at 25°C, which compares favorably with the rates of 2.34 and 2.5 Å/sec for dry oxygen- and steam-grown oxides, respectively.

As in the case of Al_2O_3 on silicon, a large surface state density was observed in the as-formed SiO_2 on silicon [28,26,112]; the distribution of surface states was also determined [26]. Again, annealing in hydrogen at 350°C for 1/2 hr reduced the surface state density to the range of 0.7 to $3 \times 10^{11}/cm^2$ eV [9,112]. However, the density of the fast surface states increased under a bias-temperature stress of 10^6 V/cm-300°C. Further annealing at 1,000°C in hydrogen or nitrogen reduced the surface state density to 0.7 to $2 \times 10^{10}/cm^2$ eV and yielded devices which were stable under bias-temperature stressing; this is no longer a low-temperature process.

M. Silver

Tiapkina and Dankov [5] have shown the oxide formed on a silver anode of an oxygen glow discharge to be of the cuprite type, Ag_2O, with a cubic lattice constant of 4.69 Å. This is the same form as the thicker thermally-grown oxide. The thin oxides formed in the plasma were found to have an irregular structure, that is, a deformation of the oxide lattice due to deviation from crystallographic conformity.

N. Tantalum

Lee et al. [63] have fitted their ellipsometric data for Ta_2O_5 growth to a two-layer model with the index of refraction N_{inner} = 1.87 and N_{outer} = 2.22. Husted et al. [54] obtained 2.1, and Leslie and Knorr [65] reported 2.21 as the value of the index of refraction of Ta_2O_5. Dielectric constant values in the range 21 to 27 [23,53,54,113] and 30 to 50 [114] have been observed. Dissipation factors fall in the range of 0.01 to 0.06 [53,54,113,114]. The breakdown strength is found to be 0.1 to 2.7×10^6 V/cm [113,114] which is lower than that obtained for Ta_2O_5 formed by aqueous anodization or reactive sputtering. Husted and Gruss [54] found the resistivity of 6,000-Å Ta_2O_5 films to be 2×10^{16} Ω cm. Flannery and Pollack [45] have reported the Ta-Ta_2O_5 barrier height to be 1.1 eV.

VII. SUMMARY

This chapter has reviewed the types of discharges appropriate for anodization and their properties which are unique to oxygen, and has pointed out the difficulties encountered in characterizing oxygen plasmas with the usual probe measurements. These descriptions were used to provide consistent explanations of the effects observed during anodization in the different types of oxygen discharges.

Regardless of which mechanisms are operative one must eventually answer the question: how useful are the oxides fabricated in discharges? The various techniques developed for gas anodization do not all have the same range of applicability. Microwave oxidation is limited to metals or semiconductors which cannot be thermally oxidized at or above 400°C, and has been used successfully for the high-speed oxidation of silicon. The temperature of the less dense induced RF and DC discharges is in the range of 25 to 100°C, making these techniques attractive for the oxidation of a different class of materials of low growth rates.

The quality of the oxides grown in oxygen discharges is variable, depending upon the technique and the care used in its implementation. In some cases the DC techniques produced films with significant sputter contamination, while the DC arc and microwave methods produced silicon dioxide whose physical and electrical properties were equal to that of the best thermally grown oxide. In order to achieve the low surface state density characteristic of a high quality oxide, these films must first be annealed at 1,000°C in hydrogen; this, however, completely eliminates any advantage gained by growing the oxide at low temperatures.

Both Al_2O_3 and SiO_2 grown at low temperatures and annealed

TABLE 2

A Summary of Publications on Plasma Anodization Classified by Material and Type of Discharge and Denoted by Reference Number

Material formed	DC discharge[a]								AC discharge[b]			
	Floating	Externally applied sample voltage							Floating		Ext. voltage	
	NG	NG	FDS	PC	AG	Anode	Cathode	Arc	RF	μwave	RF	μwave
Ag_2O						5,115	115					
AlN	77,78,79,80											
Al_2O_3	46,48,62,87 88,89,90,93 116,117,118 119,132	49,56,71,72, 84,86,91,95, 96,97,98,99, 101,104,105, 106,107,120, 121,122,123				3,39 41	3,39 41				7	
CuO, Cu_2O						41	41					
Fe_2O_3						41	41					
GaAs		75							108	29		
$GaAs_xP_{1-x}$									108			
Garnet		109										
GeO_2		69				40	40					
H_fO_2		84										
$La_2Ti_2O_7$		124										

(continued)

TABLE 2 (continued)

Material formed	DC discharge[a]								AC discharge[b]			
	Floating	Externally applied sample voltage							Floating		Ext. voltage	
	NG	NG	FDS	PC	AG	Anode	Cathode	Arc	RF	μwave	RF	μwave
MgO						41	41					
Nb_2O_5	110,125	63,73			54,55						23,42	
NiO	126					39	39					
PbO	47										42	
SiO_2		50,105,131	50	50		40	44	9,112		8,27	26,58, 130	8,27, 28
Ta_2O_5	45	43,63,65,67, 73,113,114, 123,127	43		51,53 54		43	57			7,23, 24	
TiO_2		123			54							
WO					54							
WC						82						
VO_2	130				54							
ZnO					4	41	41,128					
ZrO_2					54,70, 129							

[a] NG = negative glow, FDS = Faraday dark space, PC = positive column, AG = anode glow.

[b] RF = radio frequency, μwave = microwave.

at 350°C have been found to produce stable drift-free oxides suitable for MOS or TFT applications, where bias-temperature stress will not be encountered; plasma-grown Al_2O_3 on silicon has yielded MOS transistors whose stability under high-energy electron bombardment is an order of magnitude better than that of devices produced by any other method. Oxidation at the cathode of an RF glow discharge has demonstrated that lead and indium films can be oxidized with the run-to-run uniformity necessary for Josephson junction production.

In conclusion, plasma-grown films have not had any impact on silicon transistor technology except in special circumstances such as radiation-resistant devices for space applications. Plasma techniques have eliminated the basic instability problem with TFTs --the evaporated SiO--and have made Josephson junctions a commercial possibility. Potential applications in other technologies have yet to be thoroughly explored.

RECENT DEVELOPMENT

Since completion of this chapter, two papers which deal with rate limiting steps in glow discharge oxidation have been published:

V. A. Lavrenko, A. P. Pomytkin, and E. S. Lugovoskaya, *Oxidation of Metals*, *10*, 97 (1976).

M. Popp, L. Young, D. L. Pulfrey, and G. Olive, *J. Electrochem. Soc.*, *124*, June 1977.

REFERENCES

1. E. Olsen and V. W. Meloche, *J. Amer. Chem. Soc.*, *58*, 2511 (1936).
2. A. Gunterschultze and H. Betz, *Z. Electrochem. U. Angew Phys. Chem.*, *44*, 248 (1938).
3. P. D. Dankov and D. V. Ignatov, *Dolk. Akad. Nauk SSSR*, *54*, 235 (1946).
4. D. V. Ignatov, *Dolk. Akad. Nauk SSSR*, *54*, 329 (1946).
5. V. V. Tiapkina and P. D. Dankov, *Dolk. Akad. Nauk SSSR*, *54*, 415 (1946).
6. J. L. Miles and P. H. Smith, *J. Electrochem. Soc.*, *110*, 1240 (1963).
7. P. O. Worledge and D. White, *Brit. J. Appl. Phys.*, *18*, 1337 (1967).

8. J. R. Ligenza, *J. Appl. Phys.*, *36*, 2703 (1965).

9. J. R. Ligenza and M. Kuhn, *Solid State Technology*, Dec. 1970, p. 33.

10. J. D. Cobine, *Gaseous Conductors*, Dover Publications, 1958, Chap. 8.

11. G. Francis, *Handbuch der Physik*, *22* (S. Flugge, ed.), Springer-Verlag, Berlin, 1956, p. 53.

12. J. B. Thompson, *Proc. Roy. Soc.* *262A*, 519 (1961).

13. R. W. Lunt and A. H. Gregg, *Trans. Faraday Soc.*, *36-2*, 1062 (1940).

14. H. Drost, H. D. Klotz, U. Timm, and H. Pupke, *Expt. Tech. der Physik*, *14*, 47 (1966).

15. J. B. Thompson, *Proc. Phys. Soc.* (London), *73*, 818 (1959).

16. W. S. Whitlock and J. E. Bounden, *Proc. Phys. Soc.* (London), *77*, 845 (1961).

17. H. W. Rundle, K. A. Gillespie, R. M. Yealland, R. Sova, and J. M. Deckers, *Can. J. Chem.*, *44*, 2995 (1966).

18. J. T. Herron and H. I. Schiff, *Can. J. Chem.*, *36*, 1159 (1958).

19. J. R. Ligenza and E. I. Povilonis, U.S. Patent 3,476,971.

20. G. Francis, *Ionization Phenomena in Gases*, Academic Press, New York, 1960, p. 81.

21. E. W. McDaniel, *Collision Phenomena in Ionized Gases*, Wiley, New York, 1969, p. 115.

22. A. V. Phelps, *Chemical Reactions in Electric Discharges*, Advances in Chemistry Series 80, Amer. Chem. Soc., Washington, D.C., 1969, p. 18.

23. V. S. Mikhalkin and L. L. Odynets, *Elektrokhim.*, *6*, 359 (1970).

24. V. S. Mikhalkin and L. L. Odynets, *Elektrokhim.*, *7*, 848 (1971).

25. E. V. Makara, V. S. Mikhalkin, and L. L. Odynets, *Elektrokhim.*, *7*, 1096 (1971).

26. D. L. Pulfrey and J. J. H. Reche, *Solid State Electronics*, *17*, 627 (1974).

27. J. Kraitchman, *J. Appl. Phys.*, *18*, 4323 (1967).

28. E. R. Skelt and G. M. Howels, *Surface Sci.*, *7*, 490 (1967).

29. O. A. Weinreich, *J. Appl. Phys.*, *37*, 2924 (1966).

30. L. Elias, E. A. Ogryzlo, and H. I. Schiff, *Can. J. Chem.*, *37*, 1680 (1959).

31. F. Kaufman, in *Chemical Reactions in Electric Discharges*, Advances in Chemistry Series 80, Amer. Chem. Soc., Washington, D.C., 1969, Chap. 2.

32. J. F. Waymouth, *Phys. of Fluids*, *7*, 1843 (1964).

33. L. B. Loeb, *Basic Processes of Gaseous Electronics*, University of California Press, Los Angeles, 1955, p. 361.

34. A. von Engel and M. Steenbeck *Elecktrische Gasentladungen*, Vol. 2, Springer, Berlin, 1934, Sections 14-16.

35. I. Langmuir and K. T. Compton, "Electrical Discharges--Part II," *Rev. Mod. Phys.*, 1931, Chap. III, p. 214.

36. R. F. L. Boyd and J. B. Thompson, *Proc. Roy. Soc.*, *2524*, 102 (1969).

37. J. A. Nilson and D. H. McKay, RCA Technical Report No. MNLD-71-TR-003 (1971).

38. P. D. Dankov, *Dolk. Akad. Nauk SSSR*, *51*, 449 (1946).

39. D. V. Ignatov, *J. Chim. Phys.*, *54*, 96 (1957).

40. R. I. Nazarova, *Russian J. Phys. Chemistry*, *36*, 522 (1952).

41. R. I. Nazarova, *Zh. Fiz. Khim*, *32*, 79 (1958).

42. J. Greiner, *J. Appl. Phys.*, *42*, 5151 (1971).

43. M. Scharfe, M.S. Thesis, University of Minnesota, 1966.

44. K. Z. Lertes, *Angew. Phys.*, *24*, 147 (1968).

45. W. E. Flannery and S. R. Pollack, *J. Appl. Phys.*, *24*, 4417 (1966).

46. S. Shapiro, *Phys. Rev. Lett.*, *11*, 80 (1963).

47. W. Schroen, *J. Appl. Phys.*, *39*, 2671 (1968).

48. S. R. Pollack and C. E. Morris, *Trans. AIME*, *233*, 497 (1965).

49. A. C. Nyce, Ph.D. Dissertation, University of Maryland, 1971.

50. M. A. Copeland and R. Pappu, *Appl. Phys. Lett.*, *19*, 199 (1971).

51. T. A. Jennings, W. McNeill, and R. E. Salomon, *J. Electrochem. Soc.*, *114*, 1134 (1967).

52. T. A. Jennings and W. McNeill, *Appl. Phys. Lett.*, *12*, 25 (1968).

53. T. A. Jennings and W. McNeill, The Electrochemical Society Spring Meeting, New York, May 4-9, 1969, Abstract 23.

54. D. Husted, L. Gruss, and T. Mackus, *J. Electrochem. Soc.*, *118*, 1989 (1971).

55. J. F. O'Hanlon and M. Sampogna, *J. Vac. Sci. Technol.*, *10*, 450 (1973).

56. J. F. O'Hanlon, *J. Electrochem. Soc.*, *118*, 270 (1971).

57. W. B. Orcutt, M.S. Thesis, Air Force Institute of Technology, 1972.

58. D. L. Pulfrey, F. G. M. Hawthorn, and L. Young, *J. Electrochem. Soc.*, *120*, 1529 (1973).

59. L. Young and F. G. R. Zobel, *J. Electrochem. Soc.*, *113*, 227 (1966).

60. E. O. Johnson and L. Malter, *Phys. Rev.*, *80*, 58 (1950).

61. C. J. Dell'Oca, D. L. Pulfrey, and L. Young, in *Physics of Thin Films* (M. H. Francombe and R. W. Hoffman, eds.), Academic Press, New York, 1971, p. 2.

62. E. E. Huber, Jr., F. L. Johnston, Jr., and C. T. Kirk, Jr., *J. Appl. Phys.*, *39*, 5104 (1968).

63. W. L. Lee, G. Olive, D. L. Pulfrey, and L. Young, *J. Electrochem. Soc.*, *117*, 1172 (1970).

64. J. F. O'Hanlon, *J. Vac. Sci. Technol.*, *7*, 330 (1970).

65. J. D. Leslie and K. Knorr, *J. Electrochem. Soc.*, *121*, 263 (1974).

66. G. J. Tibol and R. W. Hull, *J. Electrochem. Soc.*, *111*, 1368 (1964).

67. N. F. Jackson, *J. Mater. Sci.*, *2*, 12 (1967).

68. G. D. Olive, D. L. Pulfrey, and L. Young, *J. Electrochem. Soc.*, *117*, 945 (1970).

69. J. F. O'Hanlon, *Appl. Phys. Lett.*, *14*, 127 (1969).

70. N. Ramasubramanian, *J. Electrochem. Soc.*, *117*, 949 (1970).

71. J. Vrba and S. B. Woods, *Can. J. Phys.*, *50*, 548 (1971).

72. L. Locker and L. Skolnick, *Appl. Phys. Lett.*, *12*, 396 (1968).

73. K. Knorr and J. D. Leslie, *J. Electrochem. Soc.*, *121*, 805 (1974).

74. I. Franz and W. Langheinrich, *Solid State Electronics*, *11*, 59 (1968).

75. D. L. Pulfrey and L. Young, USAF Technical Report, AFAL-TR-69-318 (1969).

76. C. J. Dell'Oca and L. Young, *J. Electrochem. Soc.*, *117*, 1545 (1970).

77. G. Lewicki and C. A. Mead, *J. Phys. Chem. Solids*, *29*, 1255 (1968).

78. G. Lewicki and C. A. Mead, *Phys. Rev. Lett.*, *16*, 939 (1966).

79. G. Lewicki and J. Maserjian, *Trans. AIME*, *2*, 673 (1971).

80. Y. Uemura, K. Tanaka, and M. Iwate, *Thin Solid Films*, *20*, 11 (1974).

81. R. Freiser, *J. Electrochem. Soc.*, *115*, 1092 (1968).

82. H. Schmellenmeier, *Exp. Technik der Physik*, *2*, 4 (1953).

83. G. Olive, D. L. Pulfrey, and L. Young, *Thin Solid Films*, *12*, 427 (1972).

84. P. E. Norris and K. H. Zaininger, Final Report, Contract No. N00039-69-C-0540, Naval Electronic Systems Command, Feb. 1970.

85. A. Waxman and K. H. Zaininger, *Proc. IEEE*, *57*, 9 (1969).

86. R. Morgan, *J. Mater. Sci.*, *6*, 1227 (1971).

87. T. E. Hartman, *J. Appl. Phys.*, *35*, 3283 (1964).

88. W. Gericke and W. Ludwig, *phys. stat. sol. (a)*, *1*, 189 (1970).

89. K. Gundlach and J. Holzl, *Surface Sci.*, *27*, 125 (1971).

90. B. Korneffel and W. Ludwig, *phys. stat. sol. (a)*, *8*, 149 (1971).

91. A. M. Goodman, *J. Appl. Phys.*, *41*, 2176 (1970).

92. G. W. Lewicki and C. A. Mead, *Appl. Phys. Lett.*, *8*, 98 (1966).

93. K. H. Gundlach and J. Kadlec, *Appl. Phys. Lett.*, *20*, 445 (1972).

94. T. Miyazaki, T. Edahiro, and J. Naki, *Oyo Butsuri*, *36*, 797 (1967).

95. A. Waxman and K. H. Zaininger, *Appl. Phys. Lett.*, *12*, 109 (1968).

96. F. B. Micheletti, P. E. Norris, and K. H. Zaininger, Final Report No. F04701-70 C-0077, Air Force Systems Command, October 1970.

97. L. D. Locker and L. P. Skolnick, The Electrochemical Society Fall Meeting, Montreal, October 6, 1968, Abstract 483.

98. G. J. Tibol and W. M. Kaufman, *Proc. IEEE*, *52*, 1465 (1964).

99. P. E. Norris, The Electrochemical Society Spring Meeting, Washington, D.C., May 9, 1971, Abstract 7.

100. L. Young, *Anodic Oxide Films*, Academic Press, New York, 1961, p. 212-213.

101. L. E. Terry and E. E. Komarek, AEC Report No. SC-TM-307-63 (1964).

102. W. S. Nicol, *Proc. IEEE*, *56*, 109 (1968).

103. M. T. Duffy and A. G. Revesz, *J. Electrochem. Soc.*, *117*, 372 (1970).

104. A. Waxman and G. Mark, *Solid State Electronics*, *12*, 751 (1969).

105. R. V. Pappu and A. R. Boothroyd, *Appl. Phys. Lett.*, *22*, 72 (1973).

106. F. B. Micheletti and K. H. Zaininger, *IEEE Trans. Nucl. Sci.*, *NS-17*, 27 (1970).

107. K. H. Zaininger and A. S. Waxman, *IEEE Trans. Electron Devices*, *ED-16*, 333 (1969).

108. T. Sugano and Y. Mori, *J. Electrochem. Soc.*, *121*, 113 (1974).

109. A. W. Simpson and J. M. Lucas, *Proc. Brit. Ceram. Soc.*, *18*, 117 (1970).

110. L. O. Mullen and D. B. Sullivan, *J. Appl. Phys.*, *40*, 2115 (1969).

111. K. C. Park, M. Berkenblit, D. J. Herrel, T. B. Light, and A. Reisman, *J. Electr. Materials*, *2*, 201 (1973).

112. J. R. Ligenza, The Electrochemical Society Spring Meeting, Washington, D.C., May 9, 1971, Abstract 6.

113. M. C. Johnson, *Proc. IEEE*, *CP-2*, 1 (1964).

114. F. Vratny, *J. Amer. Ceram. Soc.*, *50*, 283 (1967).

115. V. V. Tiapkina and P. D. Dankov, *Dokl. Akad. Nauk SSSR*, *59*, 1313 (1948).

116. K. H. Gundlach and J. Antula, *Thin Solid Films*, *11*, 401 (1972).

117. J. Lambe and R. C. Jaklevic, *Phys. Rev.*, *165*, 821 (1968).

118. G. Lecoy and L. Gouskov, *phys. stat. sol.*, *30*, 9 (1968).

119. A. Braunstein, M. Braunstein, G. S. Picus, and C. A. Mead, *Phys. Rev.*, *14*, 219 (1965).

120. G. D. Slawecki, Ph.D. Dissertation, University of Maryland, 1969.

121. P. J. Fopiano, *IEEE Trans. PMP*, *1*, S-217 (1965).

122. A. Waxman, *Electronics*, March 18, 1968, p. 88.

123. G. J. Tibol, *Insulation*, June 1965, p. 25.

124. R. E. Whitmore and J. L. Vossen, *IEEE Trans. PMP*, *1*, S-10 (1965).

125. R. Graffe and T. Wiik, *J. Appl. Phys.*, *42*, 2146 (1971).

126. J. G. Adler and T. T. Chen, *Solid State Comm.*, *9*, 501 (1971)

127. K. Asano, M.S. Thesis, University of Minnesota, 1964.

128. K. Germey and H. Schmellenmeier, *Exp. Tech. Phys.*, *3*, 109 (1955).

129. N. Ramasubramanian, *J. Electrochem. Soc.*, *117*, 947 (1970).

130. H. Seifarth and W. Rentsch, *phys. stat. sol. (a)*, *18*, 135 (1973).

131. J. J. H. Reche, M.S. Thesis, University of British Columbia, 1973.

132. J. L. Miles, P. H. Smith, and W. Schoenbein, *Proc. IEEE*, *51*, 937 (1963).

ADDENDUM TO CHAPTER 1

213. M. D. Goldshtein, T. I. Zalkind, and V. I. Veselovskii, *Elektrokhim.*, *10*, 1533 (1974).

214. P. R. Norton, *J. Catal.*, *36*, 211 (1975).

215. P. R. Norton, *Surface Sci.*, *47*, 98 (1975).

216. M. Fujihira and T. Kuwana, *Electrochim. Acta*, *20*, 565 (1975).

217. J. P. Hoare, *Electrochim. Acta*, *20*, 267 (1975).

218. S. Shibata and M. P. Sumino, *Electrochim. Acta*, *20*, 739 (1975).

219. Y. V. Battalova, L. A. Smirnova, G. F. Volodin, and Y. M. Tyurin, *Elektrokhim.*, *11*, 1276 (1975).

220. A. J. Appleby, *J. Electroanal. Chem.*, *68*, 45 (1975).

221. R. M. Ishikawa and A. T. Hubbard, *J. Electroanal. Chem.*, *69*, 317 (1976).

222. J. L. Ord, D. J. De Smet, and M. A. Hopper, *J. Electrochem. Soc.*, *123*, 1352 (1976).

223. J. V. Dobson, T. Dickinson, and P. R. Snodin, *J. Electroanal. Chem.*, *69*, 215 (1976).

224. D. Gilroy, *J. Electroanal. Chem.*, *71*, 257 (1976).

225. A. Ward, A. Damjanovic, E. Gray, and M. O'Jea, *J. Electrochem. Soc.*, *123*, 1599 (1976).

226. C. G. Rader and B. V. Tilak, *J. Electrochem. Soc.*, *123*, 1708 (1976).

227. J. P. Hoare, R. F. Paluch, and S. G. Meibuhr, *J. Electrochem. Soc.*, *123*, 1821 (1976).

228. J. O. Zerbino, N. R. De Tacconi, A. J. Calandra, and A. J. Arvia, *J. Electrochem. Soc.*, *124*, 475 (1977).

229. S. Evans, E. L. Evans, D. E. Parry, M. J. Tricker, M. J. Walters and J. M. Thomas, *Faraday Discuss. Chem. Soc.*, *58*, 97 (1974).

230. I. N. Sorokin, A. P. Alekin, and V. P. Lavrishchev, *Russian J. Phys. Chem.*, *49*, 1697 (1975).

231. M. M. Lohrengel and J. W. Schultze, *Electrochim. Acta*, *21*, 957 (1976).

232. J. W. Schultze and M. M. Lohrengel, *Ber. Bunsenges. Phys. Chem.*, *80*, 552 (1976).

233. M. Sotto, *J. Electroanal. Chem.*, *69*, 229 (1976).

234. M. Sotto, *J. Electroanal. Chem.*, *70*, 291 (1976).

235. M. Sotto, *J. Electroanal. Chem.*, *72*, 287 (1976).

236. D. A. J. Rand, *Anal. Chem.*, *47*, 1481 (1975).

237. I. Morcos, *J. Electrochem. Soc.*, *124*, 13 (1977).

238. D. N. Buckley and L. D. Burke, *J. Chem. Soc.*, Faraday Trans. I, *71*, 1447 (1975).

239. D. N. Buckley and L. D. Burke, *J. Chem. Soc.*, Faraday Trans. I, *72*, 2431 (1976).

240. D. N. Buckley, L. D. Burke, and J. K. Mulcahy, *J. Chem. Soc.*, Faraday Trans. I, *72*, 1896 (1976).

241. V. I. Bystrov and O. P. Romashin, *Elektrokhim.*, *11*, 1226 (1975).

242. D. Galizzioli, F. Tantardini, and S. Trasatti, *J. Appl. Electrochem.*, *5*, 203 (1975).

243. L. D. Burke and J. K. Mulcahy, *J. Electroanal. Chem.*, *73*, 207 (1976).

244. V. S. Bagotskii, A. M. Skundin, and E. K. Tuseeva, *Electrochim. Acta*, *21*, 29 (1976).

AUTHOR INDEX

Numbers in brackets are reference numbers and indicate that an author's work is referred to although his or her name is not cited in text. Underlined numbers give the page on which the complete reference is listed.

SUBJECT INDEX